Math Mammoth
Grade 2 Answer Keys

for the complete curriculum
(Light Blue Series)

Includes answer keys to:

- Worktext part A
- Worktext part B
- Tests
- Cumulative Reviews

By Maria Miller

Contents

Math Mammoth
Grade 2-A Answer Key

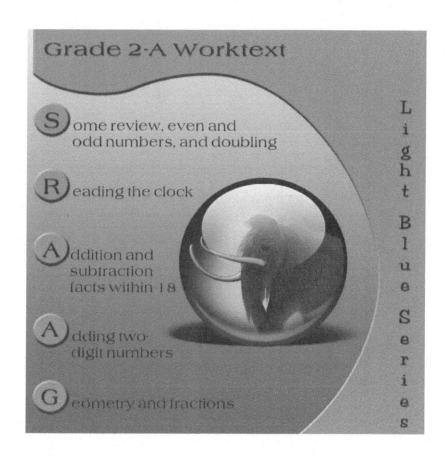

By Maria Miller

Contents

Chapter 1: Some Old, Some New

Some Review, p. 10

1. a. *32* + *7* = *39*
   b. 10 + 40 = 50
   c. 24 + 14 = 38

2. a. 35, 45, 55   b. 24, 34, 44   c. 42, 52, 62

3. a. 57, 52, 53   b. 26, 24, 25

4. a. 24 + 13 = 37   b. 28 + 40 = 68

5. a. 43   b. 31   c. 22   d. 33

6. a. 76, 59   b. 38, 59   c. 4, 7   d. 30, 40
   e. 69, 95   f. 0, 20   g. 80, 50   h. 6, 9

7. a. $15   b. 6 more.   c. 10 shirts.
   d. $22 + $20 = $42. No, I cannot buy it.

The 100-Chart and More Review, p. 12

1.

1	2	3	4	5	6	7	8	9	10
11	12	13	14	15	16	17	18	19	20
21	22	23	24	25	26	27	28	29	30
31	32	33	34	35	36	37	38	39	40
41	42	43	44	45	46	47	48	49	50
51	52	53	54	55	56	57	58	59	60
61	62	63	64	65	66	67	68	69	70
71	72	73	74	75	76	77	78	79	80
81	82	83	84	85	86	87	88	89	90
91	92	93	94	95	96	97	98	99	100

2.

1	2	3	4	5	6	7	8	9	10
11	12	13	14	15	16	17	18	19	20
21	22	23	24	25	26	27	28	29	30
31	32	33	34	35	36	37	38	39	40
41	42	43	44	45	46	47	48	49	50
51	52	53	54	55	56	57	58	59	60
61	62	63	64	65	66	67	68	69	70
71	72	73	74	75	76	77	78	79	80
81	82	83	84	85	86	87	88	89	90
91	92	93	94	95	96	97	98	99	100

3-4.

1	2	3	4	5	6	7	8	9	10
11	12	13	14	15	16	17	18	19	20
21	22	23	24	25	26	27	28	29	30
31	32	33	34	35	36	37	38	39	40
41	42	43	44	45	46	47	48	49	50
51	52	53	54	55	56	57	58	59	60
61	62	63	64	65	66	67	68	69	70
71	72	73	74	75	76	77	78	79	80
81	82	83	84	85	86	87	88	89	90
91	92	93	94	95	96	97	98	99	100

5.

1	2	3	4	5	6	7	8	9	10
11	12	13	14	15	16	17	18	19	20
21	22	23	24	25	26	27	28	29	30
31	32	33	34	35	36	37	38	39	40
41	42	43	44	45	46	47	48	49	50
51	52	53	54	55	56	57	58	59	60
61	62	63	64	65	66	67	68	69	70
71	72	73	74	75	76	77	78	79	80
81	82	83	84	85	86	87	88	89	90
91	92	93	94	95	96	97	98	99	100

6.

a. 40, 42, 44, 46, 48, 50, 52, 54, 56, 58, 60
b. 8, 18, 28, 38, 48, 58, 68, 78, 88, 98
c. 73, 71, 69, 67, 65, 63, 61, 59, 57, 55
d. 85, 80, 75, 70, 65, 60, 55, 50, 45, 40

7

a. 24 + 3 = 27	c. 33 + 6 = 39	e. 42 + 5 = 47
b. 34 + 3 = 37	d. 53 + 6 = 59	f. 32 + 5 = 37

8. a. 59 − 6 = 53   b. 39 − 6 = 33   c. 47 − 5 = 42
   d. 67 − 5 = 62   e. 60 − 3 = 57   f. 50 − 3 = 47

9.

a.	b.	c.	d.
2 + 6 = 8	4 + 4 = 8	3 + 6 = 9	8 + 2 = 10
42 + 6 = 48	74 + 4 = 78	53 + 6 = 59	48 + 2 = 50
72 + 6 = 78	94 + 4 = 98	23 + 6 = 29	98 + 2 = 100

10.

a.	b.	c.	d.
7 − 5 = 2	9 − 4 = 5	10 − 4 = 6	8 − 5 = 3
37 − 5 = 32	29 − 4 = 25	50 − 4 = 46	38 − 5 = 33
67 − 5 = 62	99 − 4 = 95	80 − 4 = 76	88 − 5 = 83

The 100-Chart and More Review, cont.

11. a. $17 + 8 = 25$ b. $35 + 6 = 41$
    c. $24 + 16 = 40$ d. $27 + 12 = 39$
    e. $19 + 24 = 43$ f. $28 + 28 = 56$

12. a. 5   b. 2   c. 2

13. a. $- 10$

50	40
52	42
64	54
23	13

b. $- 20$

100	80
20	0
40	20
21	1

c. $- 5$

45	40
95	90
96	91
11	6

Fact Families, p. 15

1. a. $6 + 5 = 11$, $5 + 6 = 11$, $11 - 5 = 6$, $11 - 6 = 5$
   b. $13 + 5 = 18$, $5 + 13 = 18$, $18 - 5 = 13$, $18 - 13 = 5$
   c. $22 + 20 = 42$, $20 + 22 = 42$, $42 - 20 = 22$, $42 - 22 = 20$

2. a. 6
   b. $3 + 7 = 10$, $7 + 3 = 10$, $10 - 7 = 3$, $10 - 3 = 7$
   c. $6 + 3 = 9$, $3 + 6 = 9$, $9 - 6 = 3$, $9 - 3 = 6$

3. a. $6 + 2 = 8$ or $2 + 6 = 8$
   b. $13 + 7 = 20$ or $7 + 13 = 20$
   c. $40 + 20 = 60$ or $20 + 40 = 60$

4. a. 9, $4 + 5 = 9$ b. 9, $2 + 7 = 9$ c. 17, $10 + 7 = 17$

5. a. 6, 100, 28   b. 87, 64, 4   c. 4, 1, 69

Puzzle corner:

a. $50 - 40 = 10$ b. $100 - 9 = 91$ c. $10 - 7 - 2 = 1$

$33 - 2 = 31$ $76 - 4 = 72$ $9 - 2 - 5 = 2$

Ordinal Numbers, p. 17

1. a.

   b.

   c. ❄❄❄❄❄❄❄❄❄❄

   d. ❄❄❄❄❄❄❄❄❄❄

e & f. E(X)T R A O R D(I)N A R Y

2. a.    b.

   c.    d.

3. BALLOONS

4. a. GRINS   b. DICTIONARY

Even and Odd Numbers, p. 19

1. a. even b. odd c. odd
   d. odd e. even f. even
   g. odd h. even i. odd

2.

Cookies	Can you share evenly?	Even or odd?
11	NO	O
14	YES	E
15	NO	O

Cookies	Can you share evenly?	Even or odd?
12	Yes	E
17	No	O
16	Yes	E

3.

1	2	3	4	5	6	7	8	9	10
11	12	13	14	15	16	17	18	19	20

21	22	23	24	25	26	27	28	29	30
31	32	33	34	35	36	37	38	39	40
41	42	43	44	45	46	47	48	49	50
51	52	53	54	55	56	57	58	59	60
61	62	63	64	65	66	67	68	69	70
71	72	73	74	75	76	77	78	79	80
81	82	83	84	85	86	87	88	89	90
91	92	93	94	95	96	97	98	99	100

4. Even numbers always end in (their last digit is)
   2, 4, 6, 8, or 0.
   Odd numbers always end in (their last digit is)
   1, 3, 5, 7, or 9.

Doubling, p. 21

1. a. $4 + 4 = 8$ b. $6 + 6 = 12$ c. $8 + 8 = 16$
   d. $10 + 10 = 20$ e. $30 + 30 = 60$ f. $50 + 50 = 100$

2.

a.
2	2
+ 2	2
4	4

b.
3	4
+ 3	4
6	8

c.
1	3
+ 1	3
2	6

d.
4	1
+ 4	1
8	2

3.

Double 1 = 2	6 + 6 = 12	11 + 11 = 22
Double 2 = 4	7 + 7 = 14	12 + 12 = 24
Double 3 = 6	8 + 8 = 16	13 + 13 = 26
Double 4 = 8	9 + 9 = 18	14 + 14 = 28
Double 5 = 10	10 + 10 = 20	15 + 15 = 30

4. Each child will make 8 sandwiches.

5. You get 6 grapes.

6. Mary moved 8 spaces. Andrea moved 12 spaces.

7. Each girl gets 4. First add $5 + 3 = 8$. Eight is double 4.

8.   13   **20**   19   **8**   15   **16**

9.

a. $8 = 4 + 4$	b. $10 = 5 + 5$	c. $4 = 2 + 2$
d. $12 = 6 + 6$	e. $14 = 7 + 7$	f. $16 = 8 + 8$

10.

3	4	5	6	7	8	9	10	11	12	13	14	15
6	8	10	12	14	16	18	20	22	24	26	28	30

11. Each woman will make 10 dolls.

12. Each teacher will get 14 worksheets.

13. You get 5 slices. First add $7 + 3 = 10$.

14. A double batch of brownies makes 32 brownies.

One-Half, p. 24

1. a.  b.

2. a.   b. c. d. e.

3.

a. $5 + 5 = 10$	b. $20 + 20 = 40$	c. $12 + 12 = 24$
$\frac{1}{2}$ of 10 is 5.	$\frac{1}{2}$ of 40 is 20.	$\frac{1}{2}$ of 24 is 12.

4.

$6 + 6 = 12$	$11 + 11 = 22$
$7 + 7 = 14$	$12 + 12 = 24$
$8 + 8 = 16$	$13 + 13 = 26$
$9 + 9 = 18$	$14 + 14 = 28$
$10 + 10 = 20$	$15 + 15 = 30$

$\frac{1}{2}$ of 16 is 8.

$\frac{1}{2}$ of 28 is 14.

$\frac{1}{2}$ of 26 is 13.

$\frac{1}{2}$ of 30 is 15.

$\frac{1}{2}$ of 22 is 11.

5.

a.	b.	c.
$\frac{1}{2}$ of 30 is 15.	$\frac{1}{2}$ of 80 is 40.	$\frac{1}{2}$ of 48 is 24.

6. a. Each one got $30.

   b. 50 students were not sick.

   c. Missie has $10 now.

   d. Mom had 20 apples in the beginning.

$10 + 10 = 20$
$15 + 15 = 30$
$20 + 20 = 40$
$25 + 25 = 50$
$30 + 30 = 60$
$35 + 35 = 70$
$40 + 40 = 80$

Adding with Whole Tens, p. 26

1. a. 64　　　b. $59 + 20 = 79$　　c. $34 + 50 = 84$
   d. $20 + 13 = 33$　　e. $40 + 26 = 66$　　f. $30 + 34 = 64$

2. a. 44　　b. $10 + 20 + 8 = 38$　　c. $20 + 20 + 4 = 44$　　d. $30 + 20 + 1 = 51$　　e. $50 + 10 + 7 = 67$
   f. $40 + 30 + 3 = 73$　　　　　　g. $60 + 20 + 3 = 83$　　h. $30 + 30 + 7 = 67$　　i. $70 + 20 + 5 = 95$

3. a. 65　b. 47, $(20 + 7 + 20)$　c. 85, $(40 + 5 + 40)$　d. 76　e. 66　f. 98　g. 97　h. 98　i. 79

4. Answers vary. For example, add first 20 and 60 to get 80. Then add 1 to that, to get 81.

5.

$5 + 5 = 10$	$30 + 30 = 60$
$10 + 10 = 20$	$35 + 35 = 70$
$15 + 15 = 30$	$40 + 40 = 80$
$20 + 20 = 40$	$45 + 45 = 90$
$25 + 25 = 50$	$50 + 50 = 100$

6. She has 15 books left to read. The student can use the chart above to find half of 30.

Adding with Whole Tens, cont.

7. The total cost is $40 + $10 + $20 = $70. Gwen paid $35. The student can use the chart above to find half of 70.

8. He has $61 – $30 = $31 left.

9. a. 10, 40, 30   b. 20, 10, 30   c. 40, 30, 70

10. What do you notice? When you add 10, 20, 30, or 40 to an even number, the answer is also an even number. When you add 10, 20, 30, or 40 to an odd number, the answer is also an odd number.

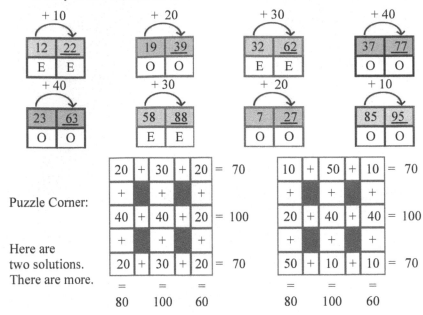

Puzzle Corner:

Here are two solutions. There are more.

Subtracting Whole Tens, p. 29

1. a. 20   b. 35   c. 26

2. a. 76, 66, 56, 46, 36, 26, 16   b. 72, 62, 52, 42, 32, 22, 12

3. a. 13, 3   b. 28, 18   c. 46, 26   d. 65, 55   e. 21, 11   f. 41, 31

4.

a. 88 – 10 = 78	b. 100 – 60 = 40	c. 34 – 10 = 24
88 – 20 = 68	90 – 50 = 40	44 – 20 = 24
88 – 30 = 58	80 – 40 = 40	54 – 30 = 24
88 – 40 = 48	70 – 30 = 40	64 – 40 = 24
88 – 50 = 38	60 – 20 = 40	74 – 50 = 24
88 – 60 = 28	50 – 10 = 40	84 – 60 = 24
88 – 70 = 18	40 – 0 = 40	94 – 70 = 24

5. a. 30 kg + 18 kg + 20 kg = 68 kg.
   b. $30 + $30 + $30 = $90. No, he cannot.
   c. The books cost $30 in total. $50 – $30 = $20. He has $20 left.

Puzzle corner: There are many solutions for both.

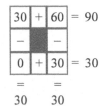

12

1.

a.	b.	c.	d.
$51 + 7 = 58$ $81 + 7 = 88$	$46 + 3 = 49$ $96 + 3 = 99$	$72 + 5 = 77$ $32 + 5 = 37$	$35 + 5 = 40$ $95 + 5 = 100$

2.

a.	b.	c.	d.
$49 - 5 = 44$ $89 - 5 = 84$	$29 - 3 = 26$ $69 - 3 = 66$	$60 - 7 = 53$ $80 - 7 = 73$	$38 - 4 = 34$ $78 - 4 = 74$

3. a. $\$20 + \$20 + \$20 = \$60$ b. The total cost was $39. $\$32 + \$1 + \$2 + 4 = \$39$.

4.

a.	b.	c.	d.
$21 + 40 = 61$ $56 + 30 = 86$	$40 + 23 = 63$ $20 + 78 = 98$	$72 - 50 = 22$ $66 - 40 = 26$	$89 - 30 = 59$ $45 - 20 = 25$

5.

W	**E**	**L**	**L**		**D**	**O**	**N**	**E**
1st	5th	9th	9th		4th	2nd	3rd	5th

6.

a.	b.	c.
$2 + 8 = 10$ $8 + 2 = 10$ $10 - 8 = 2$ $10 - 2 = 8$	$7 + 2 = 9$ $2 + 7 = 9$ $9 - 7 = 2$ $9 - 2 = 7$	$5 + 3 = 8$ $3 + 5 = 8$ $8 - 3 = 5$ $8 - 5 = 3$

7. a. $16 - 8 = 8$ b. $9 - 5 = 4$ c. $60 - 30 = 30$

8. The even numbers are 72, 60, and 8.

9.

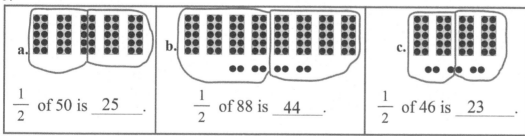

| a. $\frac{1}{2}$ of 50 is __25__. | b. $\frac{1}{2}$ of 88 is __44__. | c. $\frac{1}{2}$ of 46 is __23__. |

10. Each boy got nine cars.

11. She had 26 potatoes.

12. They have 39 colored pencils together. (Tina has 26.)

Chapter 2: Clock

Review - Whole and Half Hours, p. 37

1. a. 9 o'clock   b. 1 o'clock   c. half past 5   d. half past 11

2. a. 12 o'clock, 12:00   b. half past 12, 12:30   c. half past 6, 6:30   d. 3 o'clock, 3:00

3. a. 7:00   b. 12:30   c. 4:00   d. 3:30   e. 10:30

4. a. 5:30   b. 8:00   c. 1:00   d. 10:30   e. 2:00

The Minutes, p. 38

1. a. 25 minutes   b. 10 minutes   c. 20 minutes   d. 35 minutes

2. a. 4:15   b. 8:05   c. 3:20   d. 7:20   e. 11:25   f. 2:05   g. 3:10   h. 10:30

3. a. 2:15   b. 4:05   c. 1:20   d. 7:00   e. 9:25   f. 1:05   g. 6:10   h. 1:30

4. Clock **'c'** shows 11:25 and Clock **'a'** shows 11:05.

5. a. 9:25   b. 11:15   c. 12:20   d. 7:10

6.

	a. 1:20	b. 2:00	c. 7:15	d. 11:25
5 min. later →	1:25	2:05	7:20	11:30

	e. 10:20	f. 4:15	g. 9:00	h. 5:25
5 min. later →	10:25	4:20	9:05	5:30

The Minutes, Part 2, p. 41

1. a. 1:50   b. 2:45   c. 6:55

2.

a. 1:35   b. 2:45   c. 3:15   d. 6:55   e. 5:30   f. 7:40   g. 7:35   h. 12:20

3. a. 3:50   b. 4:40   c. 5:45   d. 11:45   e. 8:45   f. 6:35   g. 7:40   h. 7:55

4.

	a. 6:35	b. 3:40	c. 6:50	d. 9:55
5 min. later →	6:40	3:45	6:55	10:00

Puzzle corner: They started the game at 7:40.

Past and Till in Five-Minute Intervals, p. 43

1.

a. It is <u>10</u> minutes past 2 o'clock.	b. It is <u>20</u> minutes past 3 o'clock.	c. It is <u>5</u> minutes past <u>10</u> o'clock.	d. It is <u>25</u> minutes past <u>8</u> o'clock.
e. <u>20</u> past <u>1</u>	f. <u>25</u> past <u>10</u>	g. <u>10</u> past <u>7</u>	h. <u>5</u> past <u>12</u>

2.

	a. <u>15</u> past <u>5</u>	b. <u>25</u> past <u>4</u>	c. <u>5</u> past <u>9</u>
5 min. later →	<u>5:20</u>	<u>4:30</u>	<u>9:10</u>

3.

a. It is <u>10</u> minutes till 7 o'clock.	b. It is <u>5</u> minutes till <u>11</u> o'clock.	c. It is <u>25</u> minutes till <u>7</u> o'clock.	d. It is <u>20</u> minutes till <u>3</u> o'clock.
e. <u>25</u> till <u>12</u>	f. <u>15</u> till <u>8</u>	g. <u>10</u> till <u>4</u>	h. <u>5</u> till <u>3</u>
i. <u>15</u> till <u>10</u>	j. <u>5</u> till <u>6</u>	k. <u>20</u> till <u>10</u>	l. <u>25</u> till <u>2</u>

4.

a. <u>20 past 1</u>; 1:20	b. <u>15 past 12</u>; 12:15	c. <u>15 till 1</u>; 12:45
d. <u>25 till 9</u>; 8:35	e. <u>20 till 12</u>; 11:40	f. <u>20 till 6</u>; 5:40

5.

a. 8:10	b. 6:45	c. 12:25	d. 7:30
e. 9:00	f. 5:40	g. 10:55	h. 3:35

6. a. 6:45 - <u>15 till 7</u>   b. 9:30 - <u>half past 9</u>   c. 12:10 - <u>10 past 12</u>
   d. 4:55 - <u>5 till 5</u>   e. 8:35 - <u>25 till 9</u>   f. 1:40 - <u>20 till 2</u>

How Many Hours Pass?, p. 46

1.

from	5 AM	7 AM	9 AM	11 AM	10 AM
to	12 noon	1 PM	4 PM	11 PM	7 PM
hours	7	6	7	12	9

2.

Start:	8 AM	8 AM	9 AM	10 AM	8 AM
End:	12 noon	1 PM	3 PM	3 PM	2 PM
hours:	4	5	6	5	6

3.

from	4 PM	7 PM	12 noon	9 AM	7 AM
to	12 midnight	12 midnight	12 midnight	12 midnight	12 midnight
hours	8	5	12	15	17

4.

Go to bed	9 PM	8 PM	9 PM	11 PM	12 midnight
Get up	6 AM	7 AM	5 AM	9 AM	9 AM
Sleep hours	9	11	8	10	9

5. Please check the student's answer. Answers will vary.

6. a. Dad is at work for nine hours.
   b. Mary's school day is five hours long.
   c. The airplane was in the air for seven hours.

7. a. There are 24 hours in a day-night period.
   b. There are 48 hours in 2 day-night periods.

8. a. The turkey should be put into the oven at 4 p.m.
   b. He should start mowing at 11 a.m.
   c. Mom should go to bed at 11 p.m.

The Calendar: Weekdays and Months, p. 48

1. Su = Sunday, Mo = Monday, Tu = Tuesday, We = Wednesday
   Th = Thursday, Fr = Friday, Sa = Saturday

2. Monday, Tuesday, Wednesday
   Thursday, Friday, Saturday
   Saturday, Sunday, Monday

3. Answers will vary. Check the student's work.

4. a. Saturday   b. Friday   c. Wednesday   d. Monday

5. All other answers will vary other than the following:
   January 1 - Thursday   May 10 - Sunday   September 7 - Monday

6. a. Months with 31 days are: January, March, May, July, August, October and December
   b. These months have only 30 days: April, June, September, November
   c. February

7. February, March, April
   July, August, September
   October, November, December

8. a. It is three months till Anna's birthday.
   b. It is only two months till Kyle's birthday.
   c. It is six months until May's birthday.

9. Answers will vary.

1. May 5   May 12   May 19   May 26

2.

Calendar		

January
Su Mo Tu We Th Fr Sa
             ① 2
3 4 5 6 7 8 9
10 11 12 13 14 15 16
17 ⑱ 19 20 21 22 23
24 25 26 27 28 29 30
31
	February
	Su Mo Tu We Th Fr Sa
	    1 2 3 4 5 6
	7 8 9 10 11 12 13
	14 ⑮ 16 17 18 19 20
	21 22 23 24 25 26 27
	28 29
		March
		Su Mo Tu We Th Fr Sa
		      1 2 3 4 5
		6 7 8 9 10 11 12
		13 14 15 16 17 18 19
		20 ㉑ 22 23 24 ㉕ 26
		27 28 29 30 31

April
Su Mo Tu We Th Fr Sa
             1 2
3 4 5 6 7 8 9
10 11 12 13 14 15 16
17 18 19 20 21 22 23
24 25 26 27 28 29 30
	May
	Su Mo Tu We Th Fr Sa
	1 2 3 4 5 6 7
	8 9 10 11 12 13 14
	15 16 17 18 19 20 21
	22 23 24 25 26 27 28
	29 ㉚ 31
		June
		Su Mo Tu We Th Fr Sa
		          1 2 3 4
		5 6 7 8 9 10 11
		12 13 14 15 16 17 18
		19 20 21 22 23 24 25
		26 27 28 29 30

July
Su Mo Tu We Th Fr Sa
             1 2
3 ④ 5 6 7 8 9
10 11 12 13 14 15 16
17 18 19 20 21 22 23
24 25 26 27 28 29 30
31
	August
	Su Mo Tu We Th Fr Sa
	    1 2 3 4 5 6
	7 8 9 10 11 12 13
	14 15 16 17 18 19 20
	21 22 23 24 25 26 27
	28 29 30 31
		September
		Su Mo Tu We Th Fr Sa
		            1 2 3
		4 ⑤ 6 7 8 9 10
		11 12 13 14 15 16 17
		18 19 20 21 22 23 24
		25 26 27 28 29 30

October
Su Mo Tu We Th Fr Sa
                1
2 3 4 5 6 7 8
9 ⑩ 11 12 13 14 15
16 17 18 19 20 21 22
23 24 25 26 27 28 29
30 31
	November
	Su Mo Tu We Th Fr Sa
	    1 2 3 4 5
	6 7 8 9 10 ⑪ 12
	13 14 15 16 17 18 19
	20 21 22 23 ㉔ 25 26
	27 28 29 30
		December
		Su Mo Tu We Th Fr Sa
		            1 2 3
		4 5 6 7 8 9 10
		11 12 13 14 15 16 17
		18 19 20 21 ㉒ 23 ㉔
		㉕ 26 27 28 29 30 31

3. Answers will vary. Please check the student's work.

4. a. Her friend will receive the letter October 27th.

5. Julie will go back to the eye doctor in October.

6. They started playing again in January.

7.

Date 1 week ago	Date now	Date 1 week later
July 7	July 14	July 21
December 1	December 8	December 15
January 19	January 26	February 2

Date 2 weeks ago	Date now	Date 2 weeks later
July 25	August 8	August 22
October 4	October 18	November 1
February 9	February 23	March 8

8. January 26th and February 9th.

1.

a.	b.	c.	d.
1:50 10 till 2	4:25 25 past 4	8:55 5 till 9	11:05 5 past 11
e.	f.	g.	h.
3:40 20 till 4	7:25 25 past 7	5:30 half past 5	12:00 12 o'clock

2.

Time now	2:30	6:55
5 min. later	2:35	7:00

Time now	9:05	5:40
10 min. later	9:15	5:50

3. He works eight hours.

4. The class ends at 12 o'clock.

5. He goes to the chess club next time on March 24.

6. November

Chapter 3: Addition and Subtraction Facts Within 0-18

Review: Completing the Next Whole Ten, p. 59

1. a. 50, 56, 60;   d. 30, 37, 40
   b. 70, 72, 80;   e. 20, 25, 30; middle of both
   c. 90, 94, 100;   f. 30, 31, 40

2. a. 3, 3
   b. $62 + 8 = 70$, $2 + 8 = 10$
   c. $94 + 6 = 100$, $4 + 6 = 10$

3. a. $42 + 8 = 50$ b. $34 + 6 = 40$ c. $66 + 4 = 70$
   d. $61 + 9 = 70$ e. $97 + 3 = 100$ f. $83 + 7 = 90$

4. The even numbers are 8, 12, 10, and 6.

5.

a. $8 = 4 + 4$	b. $12 = 6 + 6$
c. $10 = 5 + 5$	d. $6 = 3 + 3$

6. a. 7, 8   b. 5, 6   c. 6, 7

7.

13 + 6	54 + 6	73 + 8	45 + 7	99 + 4
15 + 9	14 + 8	15 + 5	13 + 6	32 + 7
45 + 7	73 + 7	64 + 5	82 + 9	16 + 7
30 + 12	39 + 1	74 + 6	73 + 9	52 + 7
46 + 7	32 + 7	31 + 9	86 + 4	65 + 4
92 + 4	21 + 8	24 + 7	22 + 8	32 + 6
83 + 6	11 + 7	98 + 2	57 + 3	17 + 9
44 + 9	12 + 8	95 + 6	38 + 5	53 + 9
71 + 9	34 + 4	36 + 7	19 + 4	28 + 11
53 + 7	29 + 2	26 + 6	78 + 6	32 + 5

8. a. 2, 4, 3   b. 3, 5, 1   c. 3, 3, 3

9. Answers vary. For example:
   $90 + 9 + 1 = 100$; $90 + 8 + 2 = 100$; $90 + 7 + 3 = 100$;
   $90 + 6 + 4 = 100$; $90 + 5 + 5 = 100$; $90 + 4 + 6 = 100$;
   $90 + 3 + 7 = 100$; $90 + 2 + 8 = 100$; $90 + 1 + 9 = 100$

Review: Going Over Ten, p. 61

1.

a. $8 + 4 = 10 + 2 = 12$	b. $9 + 5 = 10 + 4 = 14$
c. $8 + 6 = 10 + 4 = 14$	d. $9 + 3 = 10 + 2 = 12$
e. $7 + 5 = 10 + 2 = 12$	f. $9 + 8 = 10 + 7 = 17$

2. a. $(7 + \underline{3}) + 2 = 12$ b. $(5 + \underline{5}) + 3 = 13$ c. $(8 + \underline{2}) + 4 = 14$
   d. $(6 + \underline{4}) + 4 = 14$ e. $(9 + \underline{1}) + 7 = 17$ f. $(7 + \underline{3}) + 5 = 15$

3.

a. $8 + 7$ $8 + \underline{2} + \underline{5}$ $10 + \underline{5} = \underline{15}$	b. $8 + 9$ $8 + \underline{2} + \underline{7}$ $10 + \underline{7} = \underline{17}$	c. $8 + 5$ $8 + \underline{2} + \underline{3}$ $10 + \underline{3} = \underline{13}$
d. $9 + 4$ $9 + \underline{1} + \underline{3}$ $10 + \underline{3} = \underline{13}$	e. $9 + 6$ $9 + \underline{1} + \underline{5}$ $10 + \underline{5} = \underline{15}$	f. $9 + 9$ $9 + \underline{1} + \underline{8}$ $10 + \underline{8} = \underline{18}$

4. a. 2, 3, 4   b. 3, 4, 5   c. 1, 2, 3   d. 4, 5, 6

5. The even numbers are 24, 58, 40, 100, and 2.

6. a. You have $1 left. $8 − $5 − $2 = $1.
   b. She has $15. $8 + $5 + $2 = $15
   c. Now, he has $7. $8 − $3 + $2 = $7.

Adding with 9, p. 63

1. a. 14, 14   b. $9 + 4 = 13$; $10 + 3 = 13$ c. $9 + 7 = 16$; $10 + 6 = 16$
   d. $9 + 6 = 15$; $10 + 5 = 15$ e. $9 + 8 = 17$; $10 + 7 = 17$ f. $9 + 9 = 18$; $10 + 8 = 18$

2. 4, 6, 8   10, 12, 14   16, 18, 20

3. a. 15, 15   b. 17, $10 + 7 = 17$ c. 14, $10 + 4 = 14$
   d. 16, $10 + 6 = 16$ e. 18, $10 + 8 = 18$ f. 12, $10 + 2 = 12$

4.

$9 + 0 = 9$	$9 + 5 = 14$	$9 + 9 = 18$	$9 + 4 = 13$
$9 + 3 = 12$	$9 + 6 = 15$	$9 + 1 = 10$	$9 + 10 = 19$
$9 + 7 = 16$	$9 + 8 = 17$	$9 + 2 = 11$	

5. a. 13, 17, 14 b. 16, 13, 13 c. 12, 11, 18 d. 14, 17, 15

6. a. 4, 6 b. 7, 5 c. 8, 2

Puzzle corner. a. 56, 25   b. 64, 98   c. 72, 35

Adding with 8, p. 65

1.

a. $8 + 5$	b. $8 + 4$	c. $8 + \underline{7}$
$10 + 3 = \underline{13}$	$10 + \underline{2} = \underline{12}$	$10 + \underline{5} = \underline{15}$
d. $8 + \underline{6} =$	e. $8 + \underline{8} =$	f. $8 + \underline{3} =$
$10 + \underline{4} = \underline{14}$	$10 + \underline{6} = \underline{16}$	$10 + \underline{1} = \underline{11}$

2. 4, 6, 8   10, 12, 14   16, 18, 20

$8 + 0 = 8$	$8 + 5 = 13$	$8 + 8 = 16$	$8 + 9 = 17$
$8 + 3 = 11$	$8 + 7 = 15$	$8 + 1 = 9$	$8 + 4 = 12$
$8 + 10 = 18$	$8 + 1 = 9$	$8 + 6 = 14$	$8 + 2 = 10$

3. a. 12, 14   b. 15, 13   c. 11, 17
   d. 5, 7   e. 4, 8   f. 3, 6

4. a. Jack ate 13. $8 + 5 = 13$.
   b. Eight years older. $13 - 5 = 8$ or $5 + \underline{8} = 13$.

5.

a.	b.	c.
$8 + 2 = 10$	$18 + 2 = 20$	$\frac{1}{2}$ of 0 is 0.
$8 + 4 = 12$	$18 + 4 = 22$	$\frac{1}{2}$ of 2 is 1.
$8 + 6 = 14$	$18 + 6 = 24$	$\frac{1}{2}$ of 4 is 2.
$8 + 8 = 16$	$18 + 8 = 26$	$\frac{1}{2}$ of 6 is 3.
$8 + 10 = 18$	$18 + 10 = 28$	$\frac{1}{2}$ of 8 is 4.
$8 + 12 = 20$	$18 + 12 = 30$	$\frac{1}{2}$ of 10 is 5.
$8 + 14 = 22$	$18 + 14 = 32$	$\frac{1}{2}$ of 12 is 6.

Adding with 7, p. 67

We have already studied these addition facts:

7 + 8 = 15	8 + 7 = 15
7 + 9 = 16	9 + 7 = 16
7 + 10 = 17	10 + 7 = 17

These are the new facts with 7:

7 + 4 = 11	7 + 6 = 13
7 + 5 = 12	7 + 7 = 14

1. a. 12; 13   b. 14; 15   c. 16; 17
   d. 18; 19   e. 10; 11   f. 8; 9

7 + 0 = 7	7 + 5 = 12	7 + 6 = 13	7 + 9 = 16
7 + 3 = 10	7 + 9 = 16	7 + 7 = 14	7 + 4 = 11
7 + 10 = 17	7 + 8 = 15	7 + 1 = 8	7 + 2 = 9

2. a. 11, 13, 12   b. 15, 17, 10   c. 7, 6, 8   d. 5, 9, 4
   e. 14, 16, 15   f. 11, 16, 10   g. 5, 8, 9   h. 10, 3, 5

3.

+7	
4	11
7	14
8	15
10	17
5	12
9	16

+8	
3	11
6	14
5	13
7	15
2	10
4	12

+9	
2	11
4	13
7	16
8	17
3	12
5	14

Adding with 6, p. 69

6 + 5 = 11	6 + 6 = 12

6 + 0 = 6	6 + 5 = 11	6 + 9 = 15	6 + 6 = 12
6 + 3 = 9	6 + 7 = 13	6 + 4 = 10	6 + 8 = 14
6 + 10 = 16	6 + 1 = 7	6 + 2 = 8	

1. a. 10, 12, 11   b. 14, 15, 13   c. 8, 10, 6   d. 6, 9, 5
   e. 11, 13, 10   f. 15, 14, 12   g. 7, 6, 5   h. 7, 8, 9

2.

a.	b.	c.
1 + 6 + 9 = 16	3 + 6 + 7 + 2 = 18	6 + 5 + 1 + 4 = 16
6 + 8 + 2 = 16	1 + 5 + 5 + 7 = 18	8 + 3 + 2 + 6 = 19
5 + 7 + 5 = 17	2 + 7 + 8 + 2 = 19	9 + 6 + 1 + 4 = 20

3. a. 12 apples 7 + 5 = 12
   b. It cost $6. $6 + $6 = 12 or $12 − $6 = $6.
   c. Nine bananas. 4 + 1 + 2 + 2 = 9
   d. He solved 14 problems. 9 + 5 = 14
   e. She solved 7 more problems than Ashley. 14 − 7 = 7 or 7 + 7 = 14.

Review - Facts with 6, 7, and 8, p. 71

1.

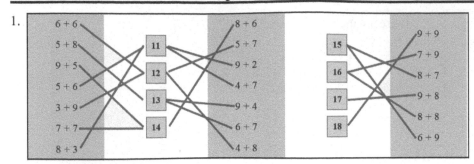

2.

a.	b.	c.
9 + <u>10</u> = 19	<u>1</u> + 16 = 17	6 + <u>0</u> = 6
8 + <u>10</u> = 18	<u>3</u> + 14 = 17	6 + <u>2</u> = 8
7 + <u>10</u> = 17	<u>5</u> + 12 = 17	6 + <u>4</u> = 10
<u>6</u> + <u>10</u> = <u>16</u>	<u>7</u> + <u>10</u> = <u>17</u>	<u>6</u> + <u>6</u> = <u>12</u>
<u>5</u> + <u>10</u> = <u>15</u>	<u>9</u> + <u>8</u> = <u>17</u>	<u>6</u> + <u>8</u> = <u>14</u>
<u>4</u> + <u>10</u> = <u>14</u>	<u>11</u> + <u>6</u> = <u>17</u>	<u>6</u> + <u>10</u> = <u>16</u>
<u>3</u> + <u>10</u> = <u>13</u>	<u>13</u> + <u>4</u> = <u>17</u>	<u>6</u> + <u>12</u> = <u>18</u>
<u>2</u> + <u>10</u> = <u>12</u>	<u>15</u> + <u>2</u> = <u>17</u>	<u>6</u> + <u>14</u> = <u>20</u>
<u>1</u> + <u>10</u> = <u>11</u>	<u>17</u> + <u>0</u> = <u>17</u>	<u>6</u> + <u>16</u> = <u>22</u>

3.

+	6	8	4	5	7	3	9
7	13	15	11	12	14	10	16
9	15	17	13	14	16	12	18
5	11	13	9	10	12	8	14

4. a. 18 elephants. 4 + 14 = 18
   b. Annie has five dolls. 10 − 5 = 5.
   c. He emptied 9 baskets. 5 + 4 = 9.
   d. She ate 17. 10 + 7 = 17

5. a. 14, 14   b. 17, 12   c. 18, 18

Subtract to Ten, p. 73

1. a. 14 − 4 = 10   b. 16 − 6 = 10   c. 15 − 5 = 10
   d. 13 − 3 = 10   e. 17 − 7 = 10   f. 19 − 9 = 10

2.

a. 15 − 7	b. 13 − 8	c. 13 − 4
/ \ 15 − 5 − 2 = 8	/ \ 13 − <u>3</u> − <u>5</u> = <u>5</u>	/ \ 13 − <u>3</u> − <u>1</u> = <u>9</u>
d. 14 − 9	e. 12 − 5	f. 16 − 8
/ \ 14 − <u>4</u> − 5 = <u>5</u>	/ \ 12 − <u>2</u> − 3 = <u>7</u>	/ \ 16 − <u>6</u> − 2 = <u>8</u>

Subtract to Ten, cont.

3.

a. $16 - 7$ $\quad\quad / \;\; \backslash$ $16 - \underline{6} - \underline{1} = \underline{9}$	b. $12 - 4$ $\quad\quad / \;\; \backslash$ $12 - \underline{2} - \underline{2} = \underline{8}$	c. $13 - 6$ $\quad\quad / \;\; \backslash$ $13 - \underline{3} - \underline{3} = \underline{7}$
d. $11 - 3$ $\quad\quad / \;\; \backslash$ $11 - \underline{1} - \underline{2} = \underline{8}$	e. $12 - 7$ $\quad\quad / \;\; \backslash$ $12 - \underline{2} - \underline{5} = \underline{5}$	f. $15 - 8$ $\quad\quad / \;\; \backslash$ $15 - \underline{5} - \underline{3} = \underline{7}$

4. a. 8, 7, 9, 6   b. 9, 6, 8, 7   c. 9, 6, 7, 8

5. a. 8, 18   b. 7, 17   c. 7, 17

Puzzle Corner:

a. $22 - 7$ $\quad\quad / \;\; \backslash$ $22 - \underline{2} - \underline{5} = \underline{15}$	b. $34 - 5$ $\quad\quad / \;\; \backslash$ $34 - \underline{4} - \underline{1} = \underline{29}$	c. $72 - 6$ $\quad\quad / \;\; \backslash$ $72 - \underline{2} - \underline{4} = \underline{66}$

Difference and How Many More, p. 75

1. a. 4   b. 4
   c. 12   d. 11

2.

The difference between 10 and 4 a. $10 - 4 = 6$	The difference between 2 and 9 b. $9 - 2 = 7$	The difference between 8 and 3 c. $8 - 3 = 5$
The difference between 20 and 50 d. $50 - 20 = 30$	The difference between 10 and 90 e. $90 - 10 = 80$	The difference between 19 and 8 f. $19 - 8 = 11$

3. a. 4   b. 2   c. 4   d. 5
   e. 1   f. 3   g. 5   h. 1

4.

a. The difference between 10 and 6 is _4_. $6 + \underline{4} = 10$	b. The difference between 6 and 12 is _6_. $6 + \underline{6} = 12$
c. The difference between 15 and 8 is _7_. $\underline{8} + \underline{7} = \underline{15}$	d. The difference between 4 and 11 is _7_. $\underline{4} + \underline{7} = \underline{11}$

5. a. 3   b. 2   c. 5

6. a. 4   b. 2
   c. 34   d. 64
   e. 20   f. 60

7. a. Jane has read three more pages. $20 - 17 = 3$ or $17 + 3 = 20$

   b. Mom has 17 eggs. $12 + 5 = 17$.

   c. She has eight pages left. $42 + 8 = 50$ or $50 - 42 = 8$.

   d. Andy worked more hours. He worked three hours more.
      Janet worked: $2 + 3 = 5$ hours. The difference: $8 - 5 = 3$.

   e. She killed 23 more flies than her husband did. $28 - 5 = 23$.

   f. She killed 19 flies. $5 + 12 + 2 = 19$.

   g. Matthew has more. He has $2 more.
      Matthew has: $12 + \$5 = \$17$. Bob has $6 + \$9 = \$15$.

Number Rainbows—11 and 12, p. 78

The student will practice mentally, and not write the answers down.
The answers are, of course, easily seen from the rainbow.
Encourage the student to also practice while keeping the rainbow covered.
There is a page for number rainbows with 13 and 14 also (later).
You can also draw or have the student draw (better) rainbows for 15, 16, 17, and 18.

Fact Families with 11, p. 80

1.

Fact families with 11		
10, 1, and 11	$10 + 1 = 11$ $\quad$ $11 - 10 = 1$	
	$1 + 10 = 11$ $\quad$ $11 - 1 = 10$	
9, 2, and 11	$9 + 2 = 11$ $\quad$ $11 - 2 = 9$	
	$2 + 9 = 11$ $\quad$ $11 - 9 = 2$	
8, 3, and 11	$8 + 3 = 11$ $\quad$ $11 - 8 = 3$	
	$3 + 8 = 11$ $\quad$ $11 - 3 = 8$	
7, 4, and 11	$7 + 4 = 11$ $\quad$ $11 - 4 = 7$	
	$4 + 7 = 11$ $\quad$ $11 - 7 = 4$	
6, 5, and 11	$6 + 5 = 11$ $\quad$ $11 - 6 = 5$	
	$5 + 6 = 11$ $\quad$ $11 - 5 = 6$	

2.

a. $11 - 10 = 1$	b. $11 - 2 = 9$	c. $11 - 3 = 8$
$\quad 11 - 9 = 2$	$\quad 11 - 4 = 7$	$\quad 11 - 6 = 5$
$\quad 11 - 6 = 5$	$\quad 11 - 5 = 6$	$\quad 11 - 9 = 2$
$\quad 11 - 8 = 3$	$\quad 11 - 7 = 4$	$\quad 11 - 4 = 7$

1.

Fact families with 12		
10, 2, and 12	$10 + 2 = 12$ $12 - 10 = 2$	
	$2 + 10 = 12$ $12 - 2 = 10$	
9, 3, and 12	$9 + 3 = 12$ $12 - 9 = 3$	
	$3 + 9 = 12$ $12 - 3 = 9$	
8, 4, and 12	$8 + 4 = 12$ $12 - 8 = 4$	
	$4 + 8 = 12$ $12 - 4 = 8$	
7, 5, and 12	$7 + 5 = 12$ $12 - 7 = 5$	
	$5 + 7 = 12$ $12 - 5 = 7$	
6, 6, and 12	$6 + 6 = 12$ $12 - 6 = 6$	

2.

a.	b.	c.	d.
$12 - 4 = 8$	$11 - 8 = 3$	$12 - 6 = 6$	$12 - 3 = 9$
$11 - 9 = 2$	$12 - 7 = 5$	$11 - 4 = 7$	$12 - 10 = 2$
$12 - 8 = 4$	$11 - 3 = 8$	$12 - 9 = 3$	$11 - 5 = 6$
$11 - 6 = 5$	$12 - 5 = 7$	$12 - 4 = 8$	$11 - 7 = 4$

3. a. 5, 3   b. 5, 4   c. 2, 4   d. 6, 3

4. Answers vary. For example: Think of the addition $8 + \_\_\_ = 11$. The number that is missing is 3 and that is the answer
to $11 - 8$.

5.

a.	b.	c.
$16 - 1 = 15$	$0 + 17 = 17$	$15 - 1 = 14$
$16 - 3 = 13$	$2 + 15 = 17$	$15 - 3 = 12$
$16 - 5 = 11$	$4 + 13 = 17$	$15 - 5 = 10$
$16 - 7 = 9$	$6 + 11 = 17$	$15 - 7 = 8$
$16 - 9 = 7$	$8 + 9 = 17$	$15 - 9 = 6$
$16 - 11 = 5$	$10 + 7 = 17$	$15 - 11 = 4$
$16 - 13 = 3$	$12 + 5 = 17$	$15 - 13 = 2$
$16 - 15 = 1$	$14 + 3 = 17$	$15 - 15 = 0$

Puzzle Corner:

a. $6 + 4 + 4 = 14$	b. $2 + 4 + 2 = 8$	c. $10 + 3 + 4 = 17$
$8 + 2 + 3 = 13$	$3 + 3 + 3 = 9$	$10 + 3 + 2 = 15$

Number Rainbows—13 and 14, p. 83

1.

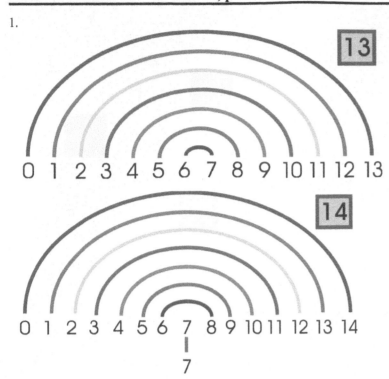

0 1 2 3 4 5 6 7 8 9 10 11 12 13

0 1 2 3 4 5 6 7 8 9 10 11 12 13 14
7

Fact Families with 13 and 14, p. 84

1.

Fact families with 13		
10, 3, and 13 ●●●●●●●●●● ●●●	10 + 3 = 13 3 + 10 = 13	13 − 10 = 3 13 − 3 = 10
9, 4, and 13 ●●●●●●●●● ●●●●	9 + 4 = 13 4 + 9 = 13	13 − 9 = 4 13 − 4 = 9
8, 5, and 13 ●●●●●●●● ●●●●●	8 + 5 = 13 5 + 8 = 13	13 − 8 = 5 13 − 5 = 8
7, 6, and 13 ●●●●●●● ●●●●●●	7 + 6 = 13 6 + 7 = 13	13 − 7 = 6 13 − 6 = 7

2.

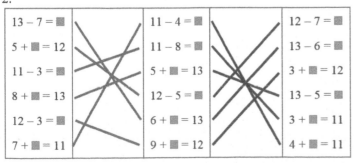

13 − 7 = ■	11 − 4 = ■	12 − 7 = ■
5 + ■ = 12	11 − 8 = ■	13 − 6 = ■
11 − 3 = ■	5 + ■ = 13	3 + ■ = 12
8 + ■ = 13	12 − 5 = ■	13 − 5 = ■
12 − 3 = ■	6 + ■ = 13	3 + ■ = 11
7 + ■ = 11	9 + ■ = 12	4 + ■ = 11

Fact Families with 13 and 14, cont.

3.

Fact families with 14		
10, 4, and 14 	$10 + 4 = 14$ $4 + 10 = 14$	$14 - 10 = 4$ $14 - 4 = 10$
9, 5, and 14 	$9 + 5 = 14$ $5 + 9 = 14$	$14 - 5 = 9$ $14 - 9 = 5$
8, 6, and 14 	$8 + 6 = 14$ $6 + 8 = 14$	$14 - 8 = 6$ $14 - 6 = 8$
7, 7, and 14 	$7 + 7 = 14$	$14 - 7 = 7$

4. a. 5, 8   b. 8, 9   c. 5, 6   d. 3, 5

5. a. 5   b. 8   c. 6
   d. 13   e. 14   f. 12
   g. 6   h. 5   i. 5

6. a. 18 cars
   b. 6 more
   c. 11 cherries
   d. 16 apples

7. a. Add 8 each time. $40 \rightarrow 48 \rightarrow 56 \rightarrow 64 \rightarrow 72 \rightarrow 80 \rightarrow 88 \rightarrow 96 \rightarrow 104$
   b. Add 4 each time. $17 \rightarrow 21 \rightarrow 25 \rightarrow 29 \rightarrow 33 \rightarrow 37 \rightarrow 41 \rightarrow 45 \rightarrow 49$

Fact Families with 15, p. 87

1.

Fact families with 15		
10, 5, and 15 	$10 + 5 = 15$ $5 + 10 = 15$	$15 - 10 = 5$ $15 - 5 = 10$
9, 6, and 15 	$9 + 6 = 15$ $6 + 9 = 15$	$15 - 6 = 9$ $15 - 9 = 6$
8, 7, and 15 	$8 + 7 = 15$ $7 + 8 = 15$	$15 - 8 = 7$ $15 - 7 = 8$

2. a. 10   b. 7   c. 11   d. 6   e. 9   g. 8

3. Answers vary: For example, think of the addition
   $9 + \_\_\_ = 15$. Because $9 + 6 = 15$, then $15 - 9 = 6$.

4. 9, 12, 15, 18, 21, 24, 27, 30, 33

5. a. Jacob has 10 peaches. $7 + 3 = 10$.
   b. Ethan has 8 books. $11 - 3 = 8$, or $8 + \_\_\_ = 11$.
   c. Noah picked 6 more. $15 - 9 = 6$ or $9 + \_\_\_ = 15$.
   d. Sophia picked 10. $15 - 5 = 10$.

6.

a. $6 = 3 + 3$	b. $12 = 6 + 6$	c. $10 = 5 + 5$
d. $18 = 9 + 9$	e. $20 = 10 + 10$	f. $8 = 4 + 4$

7. They have 8 left. $7 + 9 = 16$. Half of 16 is 8.

8. 0, 2, 4, 6, 8, 10, 12, 14, 16, 18, 20.

9.

a. bike, $28, and kite, $30 together $58	b. jeans, $47, shoes, $30, and toy $10 together $87

1.

Fact families with 16		
10, 6, and 16 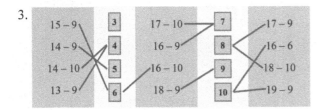	10 + 6 = 16 6 + 10 = 16	16 − 10 = 6 16 − 6 = 10
9, 7, and 16 	9 + 7 = 16 7 + 9 = 16	16 − 9 = 7 16 − 7 = 9
8, 8, and 16 	8 + 8 = 16 8 + 8 = 16	16 − 8 = 8 16 − 8 = 8

2. a. 5, 3, 6   b. 4, 7, 5   c. 6, 5, 8   d. 8, 9, 6

3.

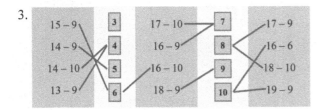

4. a.

	+ 3	+ 3	+ 3	+ 3	+ 3	+ 3	+ 3	+ 3

6   9   12   15   18   21   24   27   30

b.

	+ 4	+ 4	+ 4	+ 4	+ 4	+ 4	+ 4	+ 4

12   16   20   24   28   32   36   40   44

5. a. 20 children. 24 − 2 − 2 = 20.
   b. Yes. $10 + $4 = $14.
   c. $3. $20 − $17 = $3.
   d. $4 more. $12 + ___ = $16.
   e. More girls. There is one more girl than boys. There are five boys and six girls: 7 − 1 = 6.

6. a. >   b. <   c. >   d. >   e. >   f. <

1.

Fact families with 17		
10, 7, and 17 	10 + 7 = 17 7 + 10 = 17	17 − 10 = 7 17 − 7 = 10
9, 8, and 17 	9 + 8 = 17 8 + 9 = 17	17 − 8 = 9 17 − 9 = 8

Fact Families with 17 and 18, cont.

Fact families with 18		
10, 8, and 18 ⚫⚫⚫⚫⚫⚫⚫⚫⚫⚫ ⚫⚫⚫⚫⚫⚫⚫⚫	$10 + 8 = 18$ $8 + 10 = 18$	$18 - 8 = 10$ $18 - 10 = 8$
9, 9, and 18 ⚫⚫⚫⚫⚫⚫⚫⚫⚫ ⚫⚫⚫⚫⚫⚫⚫⚫⚫	$9 + 9 = 18$ $9 + 9 = 18$	$18 - 9 = 9$ $18 - 9 = 9$

2. a. 7, 8, 8, 9 b. 6, 7, 7, 8 c. 8, 7, 7, 6 d. 3, 4, 2, 3

3. a. > b. > c. <
   d. > e. < f. =

4. a. 6   b. 8   c. 9
   d. 15   e. 15   f. 12
   g. 8   h. 9   i. 9

5. a. The baby slept 12 hours. $4 + 2 + 3 + 3 = 12$
   b. Mom needs to buy 2 cartons of eggs. $12 + 12 = 24$, which is more than 16. She will have 8 eggs left. $24 - 16 = 8$.

6. 100 - 90 - 89 - 84 - 80 - 78 - 70

Mixed Review Chapter 3, p. 93

1. a. 2:50   b. 7:25   c. 8:55   d. 11:50

2.

	a. 2 : 35	b. 12 : 40	c. 7 : 30	d. 3 : 55
5 min. later →	2 : 40	12 : 45	7 : 35	4 : 00

3.

from	2:25	2:20	7:00	11:30	6:05
to	2:35	2:40	7:15	11:50	6:15
minutes	<u>10 minutes</u>	20 minutes	15 minutes	20 minutes	10 minutes

4. Each child got 9 raisins and 6 almonds.

5. a. $10 = 5 + 5$ b. $16 = 8 + 8$ c. $40 = 20 + 20$

6.

```
      78        71        65        60        50
     ↗ -2 ↘ -2 ↗ -5 ↘ -1 ↗ -5 ↘ -3 ↗ -2 ↘ -7 ↗ -3 ↘ -10
  80        76        70        62        53        40
```

7. a. Isabella ate 10 strawberries. Together, they ate 30 strawberries.
   b. Kyle had $20.
   c. Jane ate 22 strawberries.
   d. Emily is 26 years older than Hannah.
   e. Ann has more cars. She has four more cars.
   f. Jim has $13.

1.

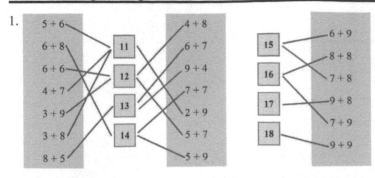

2.

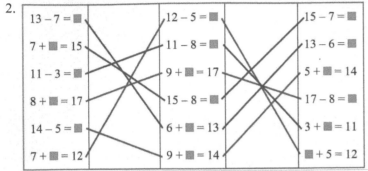

3. a. 7   b. 43   c. 7   d. 9

4. a. 7   b. 7   c. 7   d. 8   e. 6   f. 6   g. 5   h. 5   i. 8

5. 75, 70, 65, 63, 60, 54, 51

6. a.

Cookies you have	Cookies your friend has	Together we have	even/odd	Can you share evenly?
3	5	8	even	yes
5	9	14	even	yes
9	3	12	even	yes
9	7	16	even	yes

   b.

Cookies you have	Cookies your friend has	Together we have	even/odd	Can you share evenly?
5	6	11	odd	no
7	8	15	odd	no
9	4	13	odd	no
1	12	13	odd	no

7. IT GOT HOT IN THE HEAT.

8. a. Jane has seven more than Jack. $20 - 13 = 7$ or $13 + \_\_\_ = 20$
   b. Sofia has 11. $14 - 3 = 11$.
   c. Jacob has 7 pawns. $5 + 2 = 7$.
   d. You will need to save $8 more. Think: $20 + \_\_\_ = \$28$.
      After the neighbor pays you, you still need $6. You have $20 + \$2 = \$22$. Think: $22 + \underline{6} = \$28$.
   e. I need seven more squares to get to the end of the game. You roll $5 + 6 = 11$, and $11 + \underline{7} = 18$.
      To get to the end, you need to roll a total of seven on two dice. You could roll 3 and 4, or 1 and 6, or 2 and 5.

Chapter 4: Regrouping in Addition

Going Over to the Next Ten, p. 101

1. a. 22 b. 23 c. 24 d. 31 e. 31 f. 46 g. 42 h. 52 i. 63

2.

a. 28 + 8	b. 47 + 5	c. 79 + 9
28 + <u>2</u> + <u>6</u> <u>30</u> + <u>6</u> = <u>36</u>	47 + <u>3</u> + <u>2</u> <u>50</u> + <u>2</u> = <u>52</u>	79 + <u>1</u> + <u>8</u> <u>80</u> + <u>8</u> = <u>88</u>
d. 39 + 3	e. 27 + 5	f. 38 + 7
39 + <u>1</u> + <u>2</u> <u>40</u> + <u>2</u> = <u>42</u>	27 + <u>3</u> + <u>2</u> <u>30</u> + <u>2</u> = <u>32</u>	38 + <u>2</u> + <u>5</u> <u>40</u> + <u>5</u> = <u>45</u>

3. a. 13 + 9 = 22   b. 16 + 13 = 29

4. a. 19 + 7 = 26.

```
|+++++++++++++++++++++++++++++++++++++|
0        5        10        15        20        25        30
```

b. 14 + 18 = 32.

```
|+++++++++++++++++++++++++++++++++++++|
0        5        10        15        20        25        30
```

5. a. $22 + $8 = $30 or $30 − $22 = $8. Benjamin needs $8 more.
   b. $20 + $5 + $5 = $30. Natasha has $30.
   c. 28 − 8 = 20 fruit trees left to plant.
   d. 37 − 32 = 5 came later.
   e. Half of 40 is 20. 40 − 20 − 7 = 13 grapes are left.

6.

a.	b.	c.
8 + 1 = <u>9</u>	28 + 1 = <u>29</u>	78 + 1 = <u>79</u>
8 + 2 = <u>10</u>	28 + 2 = <u>30</u>	78 + 2 = <u>80</u>
8 + 3 = <u>11</u>	28 + 3 = <u>31</u>	78 + 3 = <u>81</u>
8 + 4 = <u>12</u>	28 + 4 = <u>32</u>	78 + 4 = <u>82</u>
8 + <u>5</u> = <u>13</u>	28 + <u>5</u> = <u>33</u>	78 + 5 = <u>83</u>
8 + <u>6</u> = <u>14</u>	28 + <u>6</u> = <u>34</u>	78 + 6 = <u>84</u>

What do you notice? <u>The two patterns are similar. The ones digits are the same.</u>

Add with Two-Digit Numbers Ending in 9, p. 104

1.

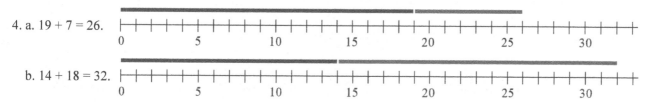

1. a. 19 + 5 = <u>24</u>	b. 29 + 7 = <u>36</u>	c. 49 + 5 = <u>54</u>
d. 29 + 8 = <u>37</u>	e. 39 + 6 = <u>45</u>	f. 49 + 9 = 58

Add with Two-Digit Numbers Ending in 9, cont.

2.

a. 19 + 7 = 26
   9 + 7 = 16
	b. 49 + 3 = 52
	   9 + 3 = 12
		c. 39 + 4 = 43
		   9 + 4 = 13

3. a. 12, 22   b. 15, 45   c. 13, 53   d. 16, 46, 36   e. 18, 78, 88   f. 14, 24, 64
   The answers to the problems in each box end in the same digit.

4. a. 5, 6, 4, 9, 8, 7   b. 13, 17, 14, 15, 18, 16   c. 7, 9, 4, 6, 5, 8   d. 15, 13, 14, 11, 17, 12

5. a. 12   b. 14   c. 11

6.

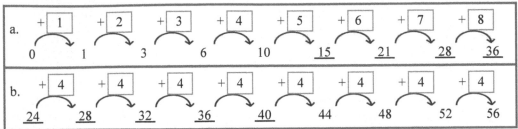

Add a Two-Digit Number and a Single-Digit Number Mentally, p. 106

1.

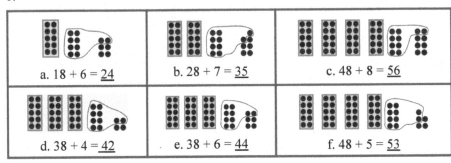

a. 18 + 6 = 24	b. 28 + 7 = 35	c. 48 + 8 = 56
d. 38 + 4 = 42	e. 38 + 6 = 44	f. 48 + 5 = 53

2. a. 18 + 7 = 25	b. 38 + 6 = 44	c. 58 + 5 = 63

3. a. 11, 21   b. 14, 44   c. 12, 82
   d. 10, 40, 30   e. 17, 77, 87   f. 13, 23, 63
   The answers to the problems in each box end in the same digit.

4. a. 13, 33   b. 14, 84   c. 14, 94
   d. 13, 43   e. 15, 35   f. 15, 55

5. To add 73 + 8, I can use the helping problem 3 + 8 = 11 . Then since
   the answer to that is 1 more than 10, the answer to 73 + 8 is 1 more than 70 .

6. 42   b. 54   c. 64

7. a. She needs 14 more eggs. 10 + 14 = 24, or 24 − 10 = 14
   b. They have eaten 20 kilograms of potatoes. 25 − 5 = 20 OR 5 + 20 = 25

Regrouping with Tens, p. 108

1. a.   33   b.   25   c.   38   d.   27   e.   36   f.   25

33	25	38	27	36	25
+ 9	+ 8	+ 9	+ 7	+ 18	+ 27
42	33	47	34	54	52

Regrouping with Tens, cont.

2. a. 13
   + 29
   ――
   42
b. 24
  + 38
  ――
  62
c. 35
  + 19
  ――
  54
d. 24
  + 47
  ――
  71
e. 44
  + 17
  ――
  61
f. 26
  + 36
  ――
  62
g. 25
  + 55
  ――
  80
h. 39
  + 35
  ――
  74

3. a. 57   b. 72   c. 91   d. 98   e. 53
   f. 93   g. 57   h. 90   i. 87   j. 95

4. a. 79   b. 87   c. 66   d. 87   e. 74

5. a. 13 + 9 + 11 = 33

   b. 27 + 16 = 43

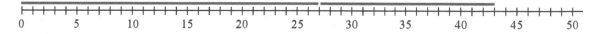

Add in Columns Practice, p. 111

1. a. 80   b. 91   c. 91   d. 70   e. 52
   f. 78   g. 116 h. 94   i. 85   j. 49

2. a. 72   b. 64   c. 72   d. 70   e. 94
   f. 62   g. 56   h. 74   i. 26   j. 66

3. a. 107   b. 142   c. 112   d. 146

4. a. 128   b. 155   c. 120   d. 147   e. 123
   f. 109   g. 126   h. 147   i. 149   j. 137

5. a. 43 hours. Add in columns 27 + 16 = 43.
   b. 13 more comic books. Subtract 29 – 16 = 13.
   c. The first vase has 2 more flowers than the second. First add 13 + 11 = 24. Then figure out the difference of 24 and 22 flowers -- it is 2 flowers.
   d. Caleb has now more money, $3 more. Add $24 + 20 = $44. Then figure out the difference of $44 and $41 -- it is $3.
   e. His bill is $43. Add $13 + $9 + $21 = $43.

Mental Addition of Two-Digit Numbers, p. 114

1.

a. 20 + 34 = 54
   20 + 30 + 4
	b. 70 + 18 = 88
	   70 + 10 + 8
		c. 50 + 27 = 77
		   50 + 20 + 7

2. a. 27, 36, 52   b. 36, 64, 87   c. 64, 83, 75

3. a. 15, 25, 45   b. 13, 23, 53   c. 12, 52, 82   d. 16, 66, 46

4.

a. 13 + 18 = 3 + 8 + 10 + 10 = 31	b. 15 + 15 = 5 + 5 + 10 + 10 = 30
c. 17 + 18 = 7 + 8 + 10 + 10 = 35	d. 19 + 15 = 9 + 5 + 10 + 10 = 34
e. 18 + 12 = 30	f. 13 + 16 = 29
g. 16 + 17 = 33	h. 17 + 15 = 32

5. a. Eight cats live outside.   b. She needs 12 more pounds of cat food.

6. 42, 45, 48, 51, 54, 57, 60, 63, 66

7.

+	2	+	4	+	6	+	8	+	10	+	12	+	14	+	16

1   3   7   13   21   31   43   57   73

8.

a. 36 + 22 30 + 20 + 6 + 2 50 + 8 = 58	b. 72 + 18 70 + 10 + 2 + 8 80 + 10 = 90
c. 54 + 37 50 + 30 + 4 + 7 80 + 11 = 91	d. 24 + 55 20 + 50 + 4 + 5 70 + 9 = 79
e. 36 + 36 30 + 30 + 6 + 6 60 + 12 = 72	f. 42 + 68 = 110
g. 45 + 18 = 63	h. 37 + 58 = 95

Puzzle corner:

a.
2	7
+ 1	4

4 1

b.
6	8
+	3

7 1

c.
2	6
+ 2	5

5 1

d.
1	3
+ 7	8

9 1

e.
3	5
+ 2	6

6 1

Adding Three or Four Numbers Mentally, p. 117

1. a. 24   b. 22   c. 22   d. 21   e. 18   f. 16

Teaching box:

7 + 5 + 3 + 6 = 10 + 11 = 21	6 + 9 + 8 + 5 = 15 + 13 = 28	9 + 8 + 8 + 4 = 16 + 4 + 9 = 29

2.

a. 8 + 8 + 2 + 8 = 16 + 10 = 26	b. 7 + 5 + 5 + 6 = 13 + 10 = 23	c. 4 + 7 + 2 + 5 = 9 + 9 = 18
d. 6 + 7 + 9 + 8 = 15 + 15 = 30	e. 8 + 5 + 2 + 6 = 10 + 11 = 21	f. 4 + 5 + 3 + 9 = 9 + 12 = 21

3. a. 18   b. 24   c. 31   d. 23   e. 20   f. 19   g. 16   h. 21   i. 22

4. Madison took 25 photos of her friends.

5. 7 + 9 = 16, so yes they can share them equally. Each boy will get 8 cars.

6. 8 + 11 = 19 which is an odd number, so no, they cannot share them equally.

7. a. 80   b. 71   c. 54   d. 47

8. a. <   b. =   c. =   d. <

1. a. 17   b. 18   c. 23

2. a. 93   b. 92   c. 98   d. 82
   e. 97   f. 96   g. 89   h. 80
   i. 107   j. 117   k. 119   l. 120

3.

$\begin{array}{r} 2 \\ 1\ 7 \\ 1\ 7 \\ +\ 4\ 9 \\ \hline 8\ 3 \end{array}$	$\begin{array}{r} 2 \\ 1\ 7 \\ 1\ 7 \\ +\ 1\ 7 \\ \hline 5\ 1 \end{array}$	$\begin{array}{r} 2 \\ 2\ 4 \\ 2\ 4 \\ 1\ 9 \\ +\ 1\ 9 \\ \hline 8\ 6 \end{array}$
a. Two shirts for $17 each; a pair of jeans for $49.	b. Three buckets for $17 each.	c. Two shovels for $24 each; two rakes for $19 each.
$\begin{array}{r} 1 \\ 8\ 9 \\ 1\ 2 \\ +\ \ 7 \\ \hline 1\ 0\ 8 \end{array}$	$\begin{array}{r} 1 \\ 3\ 6 \\ 3\ 6 \\ 2\ 3 \\ +\ 2\ 3 \\ \hline 1\ 1\ 8 \end{array}$	$\begin{array}{r} 3 \\ 1\ 9 \\ 2\ 9 \\ 2\ 9 \\ +\ 2\ 9 \\ \hline 1\ 0\ 6 \end{array}$
d. A phone for $89, a phone cover for $12, and chocolate for $7.	e. Two adult tickets for $36 each and two child tickets for $23 each.	f. A child meal for $19 and three adult meals for $29 each.

4. The person did not regroup (carry) but instead wrote the sum of the ones directly under the line, and then added the tens. The correct answers are: a. 81   b. 94

5. a. 13 more people. 35 − 22 = 13.
   b. Originally there were 6 people. 6 + 13 = 19.
   c. No, because 33 + 9 is 42, which is more than 40.
   d. You need 2 buses for 76 people because 40 + 40 = 80, and 80 is more than 76.
      You need 3 buses for 99 people, because 40 + 40 = 80 is not enough, but 40 + 40 + 40 = 120 is enough.
   e. Now there are 34 people. 40 − 6 = 34.
   f. There were 30 people left on the bus. 40 − 3 − 7 = 30.

6. a. 100   b. 110   c. 113   d. 115

7.

Number	Even?	Odd?	As a double:
8	X		4 + 4
16	X		8 + 8
100	X		50 + 50
19		X	

Number	Even?	Odd?	As a double:
18	X		9 + 9
24	X		12 + 12
15		X	
21		X	

Puzzle Corner:

1. a. 3   b. 5   c. 4   d. 12   e. 20   f. 16

2. a. 7   b. 20   c. 40   d. 8   e. 25   f. 91

3.

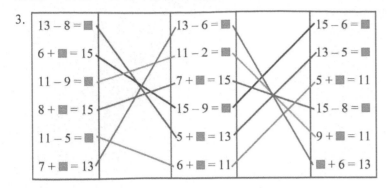

4. a. 6, 7   b. 4, 7   c. 6, 8   d. 8, 8

5. a. 3:55   b. 10:05   c. 3:45   d. 12:40

6. a. 4:05   b. 10:15   c. 3:55   d. 12:50

7. a. Shaun got the most points. Shaun got $14 + 14 = 28$ points.
   b. My brother is 8 years older than I am. $8 + 8 = 16$; $16 - 8 = 8$.
   c. Emma got 38 points. $31 + 7 = 38$.
   d. It costs $32. $26 + 6 = $32.
   e. Eight birds. $15 - 7 = 8$.
   f. Each girl gets 9 markers. $12 + 6 = 18$. Half of 18 is 9.

Review Chapter 4, p. 125

1. a. 27, 52   b. 36, 87   c. 64, 75

2. a. 35, 61   b. 27, 81   c. 92, 95

3.

a.	36 + 22	b.	72 + 18
	30 + 20 + 6 + 2		70 + 10 + 2 + 8
	__50__ + __8__ = __58__		__80__ + __10__ = __90__
c.	54 + 37	d.	24 + 55
	50 + 30 + 4 + 7		20 + 50 + 4 + 5
	__80__ + __11__ = __91__		__70__ + __9__ = __79__

4. a. Ted earned $50. $25 + $25 = $50. Together they earned $75. $50 + $25 = $75.
   b. Leah has 12. Half of 24 is 12.

5. a. 71   b. 72   c. 71   d. 61   e. 72
   f. 93   g. 95   h. 121   i. 117   j. 117

6. a. The total cost was $73. $18 + $15 + 40 = $73.
   b. The total bill is $102. $34 + $34 + $34 = $102.
   c. There are 106 stickers in total. $29 + 29 + 22 + 26 = 106$.
   d. No, he cannot. $47 + $15 = $62, which is less than $65. He needs $3 more.

Chapter 5: Geometry and Fractions

Shapes Review, p. 130

1. Check the student's pictures. It has 3 vertices.

2. Check the student's pictures. It has 4 vertices.

3. Check the student's work.

4. It has 5 vertices and 5 sides.

5. It has 6 vertices and 6 sides.

6. A circle has no vertices or straight sides.

7.

8.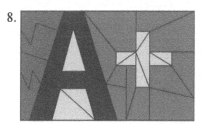

9. a. quadrilaterals   b. triangles   c. 5-sided is a pentagon and 6-sided is a hexagon

Surprises with Shapes, p. 133

1. A pentagon:

2. A triangle and a quadrilateral. You can draw the one line from corner to corner in many different ways; here is one example:

3. Answers vary as it is possible to draw these lines in many different ways. One example:

4. A hexagon:

5. & 6. Check the students' work.

7. A five-pointed star:

8. A pentagon.

9. A pentagon:

10. A six-pointed star:

11. A hexagon.

12. A hexagon:

1.

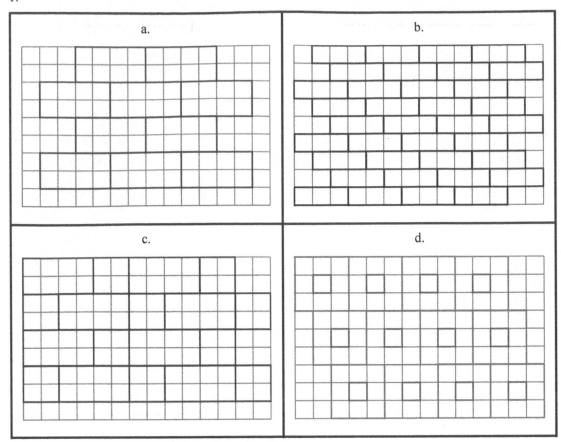

2. a. 4 little squares   b. 20 little squares   d. 16 little squares

3. a.   b.

c.   d.

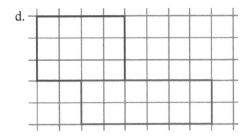

Rectangles and Squares, cont.

4.
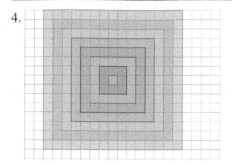

Making Shapes, p. 138

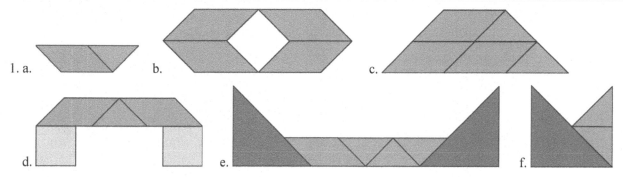

1. a.   b.   c.

d.   e.   f.

Geometric Patterns, p. 141

1.

2.

3. a. Circles and squares

b.

Solids, p. 144

1. The teacher will assist the student in making the shapes from the cut-outs.

2. a. 6 faces. They are squares.   b. 6 faces. They are rectangles.
   c. 5 faces. The bottom face is a rectangle or a square. The other four are triangles.

3. Examples: Ice cream cones, party hats, decorations on towers, a funnel.

4. a. cylinder b. cylinder c. pyramid d. cube e. cone f. box
   g. cone h. pyramid i. pyramid j. pyramid k. box l. cylinder

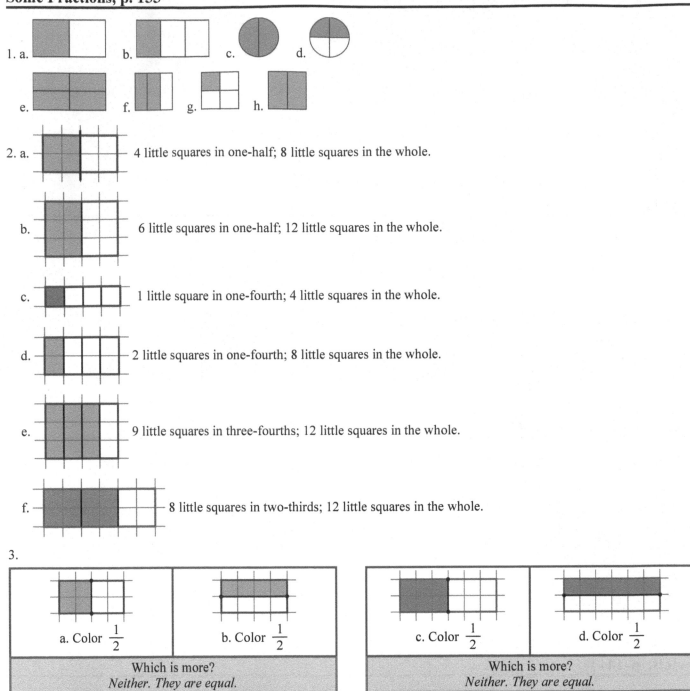

1. a. ... b. ... c. ... d.

e. ... f. ... g. ... h.

2. a. 4 little squares in one-half; 8 little squares in the whole.

b. 6 little squares in one-half; 12 little squares in the whole.

c. 1 little square in one-fourth; 4 little squares in the whole.

d. 2 little squares in one-fourth; 8 little squares in the whole.

e. 9 little squares in three-fourths; 12 little squares in the whole.

f. 8 little squares in two-thirds; 12 little squares in the whole.

3.

a. Color $\frac{1}{2}$ | b. Color $\frac{1}{2}$

Which is more?
Neither. They are equal.

c. Color $\frac{1}{2}$ | d. Color $\frac{1}{2}$

Which is more?
Neither. They are equal.

e. Color $\frac{3}{4}$ | f. Color $\frac{3}{4}$

Which is more?
Neither. They are equal.

g. Color $\frac{1}{3}$ | h. Color $\frac{1}{3}$

Which is more?
Neither. They are equal.

4. 1/2   b. 2/4   c. 2/3   d. 3/3

1.

a.	$1 = \dfrac{3}{3}$ *"One whole is 3 thirds."*	b.	$1 = \dfrac{2}{2}$ *"One whole is 2 halves."*
c.	$1 = \dfrac{4}{4}$ *"One whole is 4 fourths."*	d.	$1 = \dfrac{3}{3}$ *"One whole is 3 thirds."*

2.

a. $\dfrac{1}{3}$ < $\dfrac{1}{2}$	b. $\dfrac{2}{3}$ < $\dfrac{3}{4}$	c. 1 whole > $\dfrac{3}{4}$
d. $\dfrac{2}{4}$ = $\dfrac{1}{2}$	e. 1 whole = $\dfrac{2}{2}$	f. $\dfrac{1}{2}$ < $\dfrac{2}{3}$

3.

a. 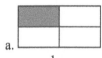 < 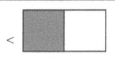 $\dfrac{1}{4}$ $\dfrac{1}{2}$	b. = $\dfrac{2}{2}$ $\dfrac{3}{3}$

4.

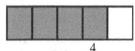 a. Color $\dfrac{4}{5}$	 b. Color $\dfrac{2}{5}$	 c. Color $\dfrac{5}{6}$	 d. Color $\dfrac{1}{6}$

5.

a. $\dfrac{1}{5}$ > $\dfrac{1}{6}$	b. $\dfrac{3}{4}$ > $\dfrac{3}{5}$	c. $\dfrac{4}{6}$ = $\dfrac{2}{3}$
d. $\dfrac{2}{5}$ < $\dfrac{1}{2}$	e. $\dfrac{3}{6}$ = $\dfrac{1}{2}$	f. $\dfrac{1}{5}$ < $\dfrac{1}{4}$

Comparing Fractions, cont.

6.

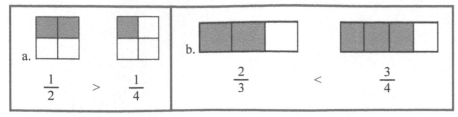

Mixed Review Chapter 5, p. 160

1. a. 5   b. 20
   c. 8   d. 52

2. a. 3   b. 5   c. 2

3. a. 9   b. 5   c. 7
   d. 12   e. 11   f. 13

4. a. 11, 21   b. 12, 92   c. 15, 55
   d. 53   e. 56   f. 93

5. a. 104   b. 97   c. 95   d. 107   e. 96

6. a. $55   b. $112   c. $116

7.

a. 8 + 8 + 2 + 8 = 16 + 10 = 26	b. 9 + 5 + 5 + 8 = 17 + 10 = 27	c. 6 + 7 + 3 + 5 = 11 + 10 = 21
d. 7 + 7 + 8 + 8 = 30	e. 9 + 4 + 4 + 7 = 24	f. 6 + 4 + 4 + 9 = 23

8. a. It costs $101. $78 + $23 = $101
   b. The first shirt costs $14 more. $29 − $15 = $14.
   c. They cost together $44. $29 + $15 = $44.

Review Chapter 5, p. 162

1. A hexagon

2. Answers vary. For example:

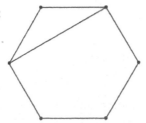

3.

4.

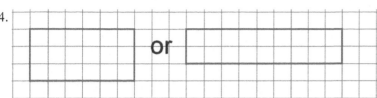

5. A cube. It has 6 faces. The faces are in the shape of a square.

6. She got a quadrilateral (to be exact, a parallelogram).

7. a. box   b. pyramid   c. cone

8.

   $1 = \dfrac{4}{4}$ $1 = \dfrac{3}{3}$

a.   b.

9.

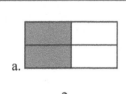

a.

$\dfrac{2}{4}$

b.

$\dfrac{1}{3}$

c.

$\dfrac{2}{3}$

d.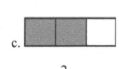

$\dfrac{2}{2}$

10.

a.

$\dfrac{1}{3}$ < $\dfrac{1}{2}$

b.

$\dfrac{2}{3}$ < $\dfrac{3}{4}$

c.

1 whole > $\dfrac{3}{4}$

Math Mammoth
Grade 2-B Answer Key

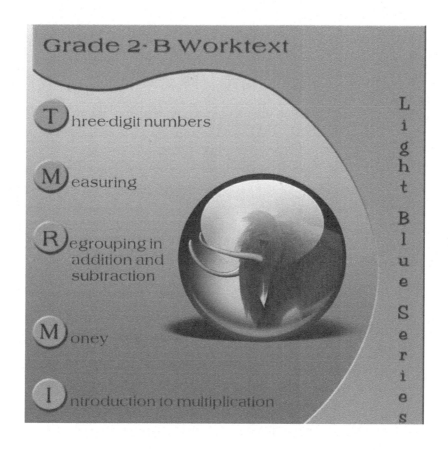

By Maria Miller

Contents

Chapter 6: Three-Digit Numbers

Three-Digit Numbers, p. 10

1.

	hundreds	tens	ones
a.	1	0	1

	hundreds	tens	ones
b.	1	0	6

	hundreds	tens	ones
c.	1	1	1

	hundreds	tens	ones
d.	1	1	3

	hundreds	tens	ones
e.	1	2	0

	hundreds	tens	ones
f.	1	2	5

	hundreds	tens	ones
g.	1	5	0

	hundreds	tens	ones
h.	1	6	2

2.

a. two hundred four

hundreds	tens	ones
2	0	4

b. two hundred thirteen

hundreds	tens	ones
2	1	3

c. three hundred twenty

hundreds	tens	ones
3	2	0

three hundred forty-five

H	T	O
3	4	5

five hundred six

T	T	O
5	0	6

five hundred forty-six

T	T	O
5	4	6

g. seven hundred ten

H	T	O
7	1	0

h. one thousand

Th	H	T	O
1	0	0	0

3.

a. 300 + 10 + 4

H	T	O
3	1	4

b. 500 + 30 + 4

H	T	O
5	3	4

c. 800 + 20 + 5

H	T	O
8	2	5

d. 800 + 60 + 4

H	T	O
8	6	4

e. 300 + 40 + 0

H	T	O
3	4	0

f. 500 + 0 + 8

H	T	O
5	0	8

Three-Digit Numbers, cont.

4.

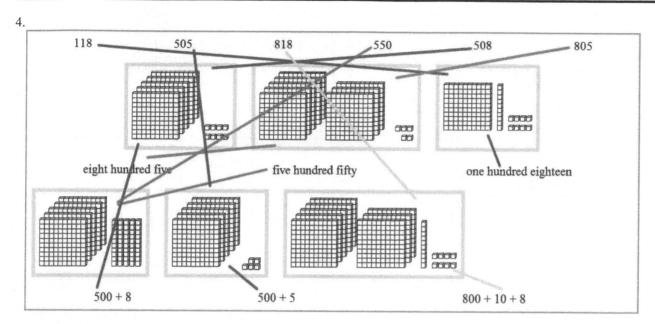

5. b. 254   c. 428   d. 550

Puzzle corner:
There are 100 tens in a thousand.

Hundreds on the Number Line, p. 14

1. a. 119, 135, 102, 129
   b. 118, 135, 130, 117
   c. 195, 180, 168 188
   d. 174, 198, 190, 167

2. a. 6   b. 7   c. 9   d. 16

3.

| 105 | 106 | 107 | 108 | 109 | 110 | 111 | 112 |

| 193 | 194 | 195 | 196 | 197 | 198 | 199 | 200 |

4.

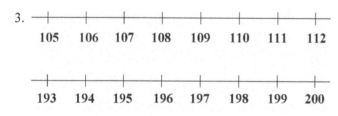

5.

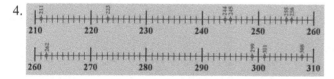

6. a. answers are in red below the line   b. answers are in purple above the line

Forming Numbers—and Breaking Them Apart, p. 16

1.

a. 276 = 2 hundreds 7 tens 6 ones = 200 + 70 + 6	b. 867 = 8 hundreds 6 tens 7 ones = 800 + 60 + 7
c. 350 = 3 hundreds 5 tens 0 ones = 300 + 50 + 0	d. 707 = 7 hundreds 0 tens 7 ones = 700 + 0 + 7
e. 409 = 400 + 0 + 9 f. 601 = 600 + 0 + 1 g. 558 = 500 + 50 + 8	h. 940 = 900 + 40 + 0 i. 383 = 300 + 80 + 3 j. 627 = 600 + 20 + 7

2. a. 733, 150   b. 245, 407

3. a. 728, 933   b. 650, 801   c. 142, 180   d. 330, 509

4. a. △ = 924   b. ▲ = 460   c. ▲ = 107

5. a. ▲ = 30   b. △ = 9   c. ▲ = 200

6. a. ▲ = 63   b. ▲ = 7   c. △ = 14

7. Answers will vary. Please check the student's work.

Puzzle Corner:

a. △ = 5   b. ▲ = 6   c. ▲ = 30

Skip-Counting by Tens, p. 18

1.

a. 248 + 10 = 258	b. 248 − 10 = 238
c. 314 + 10 = 324	d. 314 − 10 = 304
e. 551 + 20 = 571	f. 551 − 20 = 531

2.

a. 298 + 10 = 308	b. 491 + 10 = 501
c. 194 + 10 = 204	d. 398 + 10 = 408

3. a. 704, 714, 724, 734, 744, 754, 764, 774
   b. 331, 341, 351, 361, 371, 381, 391, 401
   c. 467, 477, 487, 497, 507, 517, 527, 537

4. a. 198   b. 491   c. 94   d. 299

5. a. 610, 620, 630   b. 688, 698, 708
   c. 700, 710, 720   d. 596, 606, 616
   e. 119, 129, 139   f. 495, 505, 515

6. a. 731, 721, 711, 701, 691, 681, 671, 661
   b. 950, 940, 930, 920, 910, 900, 890, 880

More-Skip Counting, p. 21

1.

971	972	973	974	975	976	977	978	979	980
981	982	983	984	985	986	987	988	989	990
991	992	993	994	995	996	997	998	999	1,000

2. a. 960, 965, 970, 975, 980, 985, 990, 995
   b. 435, 440, 445, 450, 455, 460, 465, 470
   c. 370, 375, 380, 385, 390, 395, 400, 405

3. a. 968, 970, 972, 974, 976, 978, 980, 982
   b. 490, 492, 494, 496, 498, 500, 502, 504
   c. 479, 481, 483, 485, 487, 489, 491, 493

4. a. 492   b. 301   c. 990

5. The LAST number on the chart is 1,000.

710	720	730	740	750	760	770	780	790	800
810	820	830	840	850	860	870	880	890	900
910	920	930	940	950	960	970	980	990	1,000

6. a. 699, 700, 701   b. 128, 129, 130   c. 449, 450, 451
   d. 800, 801, 802   e. 670, 671, 672   f. 998, 999, 1,000

7. a. 748, 758, 768, 778, 788, 798, 808
   b. 393, 403, 413, 423, 433, 443, 453
   c. 480, 485, 490, 495, 500, 505, 510
   d. 695, 700, 705, 710, 715, 720, 725
   e. 970, 975, 980, 985, 990, 995, 1000
   f. 500, 502, 504, 506, 508, 510, 512
   g. 695, 697, 699, 701, 703, 705, 707
   h. 421, 423, 425, 427, 429, 431, 433

Which Number Is Greater? p. 23

1. a. 145 < 154   b. 234 < 324
   c. 189 < 302   d. 513 > 315

2.

a. 120 < 110	b. 240 < 420
c. 305 < 503	d. 453 < 534

3. a. 592   b. 774   c. 949   d. 605
   e. 911   f. 482   g. 433   h. 820

Which Number Is Greater, cont.

4.

a. 159 < 300	b. 122 > 100	c. 320 < 328	d. 212 < 284
e. 200 > 190	f. 600 < 860	g. 456 < 465	h. 711 > 599
i. 780 > 500	j. 107 < 700	k. 566 < 850	l. 840 > 480

5.

a. 140 < 149 < 156	b. 357 < 573 < 750
c. 133 < 239 < 286	d. 670 < 676 < 766

6.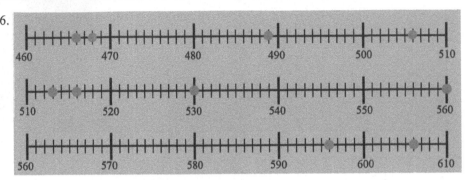

The numbers above in order are: 466 < 468 < 489 < 506 < 513 < 516 < 530 < 560 < 596 < 606

7. Answers will vary as there are many possibilities.
   a. 141, 142, ... ,148   b. 268, 269, ... , 803
   c. 280, 281, ... ,289   d. 305, 306, ... , 309

8.

100	121	127	133	167	189	200	214	212	398
145	166	134	135	120	230	212	256	347	405
156	167	137	156	155	226	356	378	380	407
632	234	138	246	267	278	476	477	450	417
432	256	200	250	245	300	355	487	478	456
355	253	289	244	305	303	570	569	490	453
361	385	377	367	356	301	537	566	505	498
689	654	390	480	478	488	675	507	508	689
654	543	489	488	483	577	589	609	504	769
723	566	570	589	578	734	631	616	789	**1000**

Comparing Numbers and Some Review, p. 26

1.

a. 150 < 515	b. 22 < 120	c. 307 < 320	d. 412 > 284
e. 240 < 750	f. 860 > 680	g. 406 < 620	h. 558 > 540
i. 605 > 450	j. 107 < 705	k. 566 < 856	l. 890 > 870

2. a. =   b. <
   c. =   d. <
   e. >   f. >
   g. >   h. <

52

Comparing Numbers and Some Review, cont.

3.
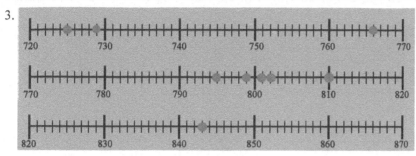

4. A SHY MILKSHAKE

5. a. 706, 670 b. 141, 411

6. Many times there are lots and lots of answers. For example, 278 > ___ + 5. You can put 0, 1, 2, 3, 4, etc. on the empty line; in fact, any number till 272 will do. Encourage the student to find other answers, and even think what are *all* the possible answers. For problems with equality (= sign) there is only one answer. For example:

a. 8, 9, 10, etc. or anything greater than 7.　　　b. 40
c. 65, 66, 67, etc. or anything greater than 64　　d. 0, 1, 2, 3, etc. or any number till 272
e. anything less than 904　　　　　　　　　　　f. any number that is 19 or more

Mystery Number:

a. 99 b. 54 c. 38

Mystery numbers: Guide the child to first think about the first hint that is given. What kind of numbers fit that hint? For example, in the first one it is the same whether read from left to right or right to left. That leaves as the only possibilities 11, 22, 33, 44, 55, 66, 77, 88, 99, and actually all one-digit numbers too. Then the child should use the second hint to choose from these.

Add and Subtract Whole Hundreds, p. 29

1. a. 200 + 100 = 300　　b. 260 + 200 = 460　　c. 250 + 100 = 350
   d. 307 + 100 = 407　　e. 262 + 200 = 462　　f. 115 + 300 = 415

2. a. 815, 578　b. 809, 663　c. 922, 694

3.

a. Cross out 100	400 − 100 ———— 300	b. Cross out 100	350 − 100 ———— 250
c. Cross out 500	736 − 500 ———— 236	d. Cross out 400	872 − 400 ———— 472

4. a. 307　b. 190
   c. 145　d. 151

5. a. 565, 248　b. 402, 380　c. 278, 94

Practice with Whole Hundreds p. 31

1. a. 700, 800   b. 602, 207   c. 340, 470

2. a. 300, 100   b. 570, 7   c. 350, 506

3. Cindy has $200 + 200 + 200 - 100 = 500$ points now.

4. a. 400, 300   b. 800, 900

5. a. 900, 700   b. 200, 300
   c. 600   d. 0

6. a. 500, 400   b. 300, 100
   c. 300, 500   d. 0, 200

7. a. They drive 250 km. $150 + 100 = 250$
   b. 500 km. $250 + 250 = 500$

8. a. $100 + 100 + 100 + 100 = 400$ cents
   b. $62 + 200 = 262$ cents
   c. 235 cents (or $2.35). Solution: $3 is 300 cents. $535 - 300 = 235$.
   d. Each girl gets 400 marbles.
   e. 300 meters
   f. 900 points. $800 - 100 + 200 = 900$

9.
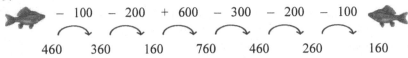

	− 100	− 200	+ 600	− 300	− 200	− 100	
460	360	160	760	460	260	160	

10. a. △ = 100   b. △ = 400   c. △ = 700
    d. △ = 400   e. △ = 900   f. △ 400

Completing the Next Hundred p. 34

1.

H | | | | | | + | | | | = H H a. 160 + 40 = 200	H H | | | | | | + | | | | | = H H H b. 250 + 50 = 300
H | | | | | | | | + | | = H H c. 180 + 20 = 200	H H | | + | | | | | | | | = H H H d. 220 + 80 = 300

2.

a.	b.	c.
$80 + \underline{20} = 100$	$30 + \underline{70} = 100$	$40 + \underline{60} = \underline{100}$
$280 + \underline{20} = 300$	$330 + \underline{70} = \underline{400}$	$740 + \underline{60} = \underline{800}$
$680 + \underline{20} = \underline{700}$	$530 + \underline{70} = \underline{600}$	$940 + \underline{60} = \underline{1000}$

3.

a.	b.	c.
$540 + \underline{60} = 600$ ($40 + 60 = 100$)	$250 + \underline{50} = 300$ ($50 + \underline{50} = 100$)	$630 + \underline{70} = 700$ ($\underline{30} + \underline{70} = 100$)
d.	e.	f.
$120 + \underline{80} = \underline{200}$ ($\underline{20} + 80 = 100$)	$440 + \underline{60} = \underline{500}$ ($\underline{40} + \underline{60} = 100$)	$970 + \underline{30} = \underline{1000}$ ($\underline{70} + \underline{30} = 100$)

54

Completing the Next Hundred, cont.

4.

a. $500 - 440 = 60$ ($440 + 60 = 500$)	b. $800 - 710 = 90$ (difference of 800 and 710)	c. $1000 - 960 = 40$ ($960 + 40 = 1000$)

5. a. 150 b. 230 c. 220 d. 330

6. a. 50, 250, 650   b. 60, 560, 160   c. 10, 710, 810

7. a. 60, 460   b. 690, 90   c. 20, 520

8. a. She will need 80 more rocks to have 200. $200 - 120 = 80$. She will need 180 more rocks to have 300.
   b. She has 120 rocks now. $100 + 30 - 10 = 120$
   c. $70 more. $200 - 130 = 70$
   d. He bought 3 windows. $400 + 200 + 200 + 200 = 1,000$.
   e. The total cost is $300. $200 + 50 + 50 = 300$.
   f. Mom spent $170 for groceries $200 - 30 = 170$.

Adding Whole Tens p. 37

1.

H |||||| and ||| a. $160 + 30 = 190$	H H H |||||| and || b. $350 + 20 = 370$
H H |||| and ||| ∙∙∙∙∙∙∙ c. $247 + 30 = 277$	H H H H	and |||||| ∙∙∙ d. $413 + 70 = 483$

2. a. 380   b. 482
   c. 589   d. 244
   e. 370   f. 197

3. a. 80, 280, 680   b. 91, 591, 891   c. 85, 285, 785

4. a. 540, 450   b. 349, 573   c. 772, 184

5. a. $232 + 40 = 272$ b. $224 + 60 = 284$ c. $423 + 40 = 463$

6.

50	100	150	200	250
300	350	400	450	500
550	600	650	700	750
800	850	900	950	1000

7. a. Roy has 40 pennies more than Rhonda, because $640 - 600 = 40$. Roy needs 60 more pennies.
   b. Jack gets 300 pennies.
   c. $640 + 30 + 30 = 700$ so Roy needs 30 pennies more.

8. a. 60, 70   b. 50, 60   c. 20, 60

Puzzle corner:

$30 + 40 - 50 = 20$ $670 - 50 + 20 = 640$

$100 - 40 - 50 = 10$ $930 - 30 + 50 = 950$

$240 + 40 - 50 = 230$ $430 + 40 + 50 = 520$

$140 - 80 + 50 + 20 = 130$ $200 + 40 + 50 - 30 = 260$

1. a. 330   b. 412

2. a. 620   b. 510
   c. 438   d. 214

3. a. 30, 130, 630   b. 40, 540, 840   c. 14, 114, 714

4. a. $272 - 50 = 222$ b. $440 - 40 = 400$ c. $343 - 20 = 323$

5. Answers will vary, but should include the idea of only subtracting in the tens, or only the tens digit changing. For example: I subtract the tens, $80 - 50 = 30$. The hundreds and ones don't change, and the answer is 633.

6. He pays $180. $100 + 80 = 180$

7. Ryan picked 215 apples. $245 - 30 = 215$

8. Janet earns $100 per week. $25 + 25 + 50 = 100$. She earns $200 in two weeks and $300 in three weeks. She will need to work four weeks to be able to buy the bicycle.

9.

a. $240 - 220 = 20$ ($220 + 20 = 240$)	b. $760 - 710 = 50$ (difference of 760 and 710)	c. $1000 - 920 = 80$ ($920 + 80 = 1000$)
d. $590 - 500 = 90$	e. $996 - 966 = 30$	f. $452 - 432 = 20$

10.

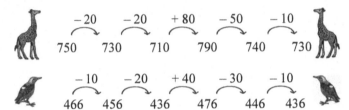

Puzzle corner:

a. Answers may vary.

$500 - 40 - 50 = 410$	$200 + 20 - 70 = 150$
$500 + 40 - 50 = 490$	$200 - 20 + 70 = 250$
$500 - 40 + 50 = 510$	$200 - 20 - 70 = 110$
$500 + 40 + 50 = 590$	$200 + 20 + 70 = 290$

b. Answers will vary.
   The smallest possible answer is $70 + 40 - 90 = 20$.
   The greatest possible answer is $90 + 70 + 40 = 200$.

1. Bill won the game with 360 points.

Jim	Dan	Bill
540	270	330
540	370	360

2. a.

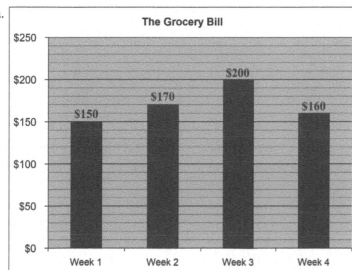

The Grocery Bill

b. They paid $40 more for week 3 than week 4. 200 − 160 = 40
c. They paid $20 more for week 2 than week 1. 170 − 150 = 20

3.

520	540	560	580	600
620	640	660	680	700
720	740	760	780	800
820	840	860	880	900
920	940	960	980	1000

4.

     −50   −30   +70   +200   −500
  700   650   620   690   890   390

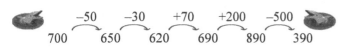

     − 100 − 30 + 80 + 300 − 30
  530   430   400   480   780   750

5.

a. 590 − 60 = 530	b. 770 + 10 = 780
590 − 70 = 520	770 + 20 = 790
590 − 80 = 510	770 + 30 = 800
590 − 90 = 500	770 + 40 = 810
590 − 100 = 490	770 + 50 = 820
590 − 110 = 480	770 + 60 = 830

Patterns and Problems, cont.

6.

a. 280 − 60 = 220	b. 590 − 80 = 510
c. 600 − 500 = 100	d. 370 − 310 = 60
e. 450 + 300 = 750	f. 716 + 60 = 776
g. 530 + 60 = 590	h. 637 + 60 = 697
i. 1000 − 300 = 700	j. 740 − 700 = 40
k. 667 − 60 = 607	l. 999 − 700 = 299

Puzzle corner:
a. 700   b. 40   c. 50   d. 90

Bar Graphs and Pictographs, p. 46

1. a. 5 students   b. 4 students   c. 7 students   d. 4 students   e. 20 students

f.	Students
Students who slept less than 8 hours	😐😐
Students who slept at least 8 hours	😐😐😐😐😐😐😐😐😐😐

2.

Page count	Number of books
200-249	2
250-299	2
300-349	6
350-399	4

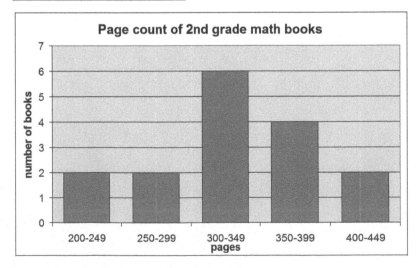

a. 4 books   b. 10 books   c. 2 books   d. 217 pages

3.
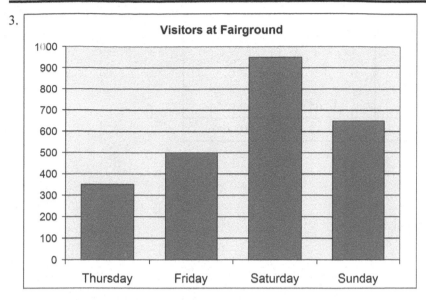

a. Saturday; 950 people
b. 650 – 500 = 150 more people
c. 350 + 500 = 850 people
d. If you don't like crowds: Thursday. If you like crowds: Saturday.

4. Check the student's work. Example: 1 ⬤ = 10 baskets

Day	Baskets
Mon	⬤⬤⬤⬤⬤⬤⬤⬤
Tue	⬤⬤⬤⬤⬤⬤
Wed	⬤⬤⬤⬤⬤⬤⬤⬤⬤
Thu	⬤⬤⬤

5.
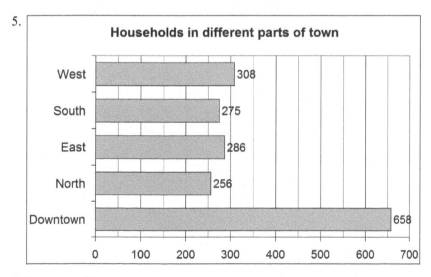

6. Answers will vary.

1.

	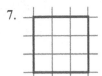 a. 2 : 15	b. 8 : 50	c. 11 : 05	d. 9 : 55
5 min. later →	2 : 20	8 : 55	11 : 10	10 : 00

2. a. 7   b. 8   c. 3
   d. 14   e. 12   f. 18

3. a. 13, 23   b. 17, 77   c. 13, 43

4. John has five plums left. John got 8 (half of 16). Then, 8 − 3 = 5.
   Jane has six plums left. Jane got 8 (half of 16). Then, 8 − 2 = 6.

5. a. 6   b. 9   c. 50
   d. 20   e. 7   f. 300

6. a. 125   b. 104   c. 104   d. 125   e. 104

7.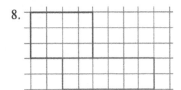

8.

9.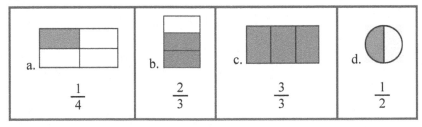

a. $\frac{1}{4}$	b. $\frac{2}{3}$	c. $\frac{3}{3}$	d. $\frac{1}{2}$

10.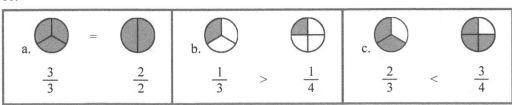

a. $\frac{3}{3} = \frac{2}{2}$	b. $\frac{1}{3} > \frac{1}{4}$	c. $\frac{2}{3} < \frac{3}{4}$

1. a. 486   b. 487   c. 496   d. 586

2. a. 178, 179, 180   b. 200, 201, 202
   c. 799, 800, 801   d. 916, 917, 918

3. a. 709   b. 674
   c. 580   d. 558

4. 695, 700, 705, 710, 715, 720

5. a. 282, 292, 302   b. 535, 545, 555

6.

200	220	240	260	280
300	320	340	360	380
400	420	440	460	480

7.

a. 238 < 265	b. 391 > 193	c. 405 < 450	d. 981 > 819
e. 8 + 600 < 60 + 800		f. 30 + 300 + 5 > 90 + 8 + 100	

8.

a. 109, 199, 901	b. 175, 177, 717

9. a. 323 + 40 = 363   b. 262 + 300 = 562

10. a. 920, 480   b. 908, 270   c. 719, 54

11. a. 600, 690   b. 929, 453   c. 542, 814

12.

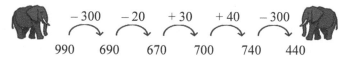

```
        − 300   − 20   + 30   + 40   − 300
      990    690    670    700    740    440
```

13. a. 260 sheep. 250 − 10 + 20 = 260
    b. 140 goldfish. 170 − 30 = 140
    c. Now he has 190 goldfish. 140 + 50 = 190. He has 100 rainbow fish. 30 + 70 = 100
    d. She traveled 860 km. 400 + 30 + 400 + 30 = 860

Puzzle corner:

a. 32   b. 8

Chapter 7: Measuring

Measuring to the Nearest Centimeter, p. 57

1. a. 7 cm   b. 3 cm
   c. 6 cm
   d. 11 cm

2. a. 8 cm   b. 6 cm

3.
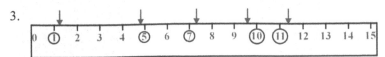

4. a. about 7 cm
   b. about 4 cm
   c. about 8 cm

5. a. about 6 cm
   b. about 2 cm
   c. about 4 cm

6. Answers will vary. Please check the students' work.

7. a. ▬▬▬▬▬▬▬▬▬

   b. ▬▬▬▬▬▬

   c. ▬▬▬▬▬▬▬▬▬▬▬▬

8. Answers will vary.

Inches and Half-Inches, p. 60

1. a. 2 inches
   b. 1 1/2 inches
   c. 2 1/2 inches
   d. 4 1/2 inches

2. a. 3 1/2 inches
   b. 1 1/2 inches
   c. 5 inches

3.
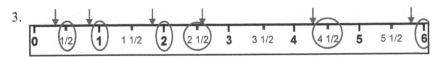

4. a. 3 inches
   b. 3 1/2 inches
   c. 4 1/2 inches
   d. 5 inches

5. a 3 1/2 inches
   b. 2 inches
   c. 5 1/2 inches

6. Please check the student's work as c. and d. are too long for the width of this page to show an example.

7. a. Rectangle, Side AB 2 1/2 inches, Side BC 1 inch, Side CD 2 1/2 inches, Side DA 1 inch.
      All the way around 7 inches.
   b. Square, Side AB 1 1/2 inches, Side BC 1 1/2 inches, Side CD 1 1/2 inches, Side DA 1 1/2 inches.
      All the way around 6 inches.

Some More Measuring, p. 63

1. a. 2 pencils   b. 4 pencils
   c. 4 pencils   d. 3 pencils
   e. The shortest pencil is 3 cm and the longest pencil
      is 11 cm. The longest pencil is 8 cm longer than
      the shortest.

2. It is a quadrilateral.
   The perimeter is about 18 cm.

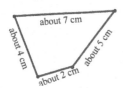

3. Answers will vary.

4. Side AB about 1 1/2 in.
   Side BC 4 in.
   Side CA 4 1/2 in.
   Perimeter about 10 in.

5. a. the centimeter-amounts
   b. one inch
   c. less than 13 inches
   d. more than 3 cm

6. Answers will vary.

Feet and Miles, p. 66

1.-4. Answers vary.

5. 1 foot is longer than 1 cm. It was about 150 cm tall.

6. It was about 2 feet tall.

7. A meter is longer. About 12 feet.

8. Answers will vary.

9. a. 390 miles   b. 120 miles

10.

Distance	Unit
from New York to Los Angeles	mi
from a house to a neighbor's house	ft
the width of a notebook	in
the distance around the earth	mi
how tall a refrigerator is	ft. in.
the width of a porch	ft
the length of a board	ft. in.

Distance	Unit
the length of a train	ft
the length of a playground	ft
from a train station to the next	mi
the width of a computer screen	in

Meters and Kilometers, p. 69

1. a. You might not be able to, but the teacher can.   b. Answers will vary.

2-4. Answers will vary.

5. a. 20 m + 20 m = 40 m
   b. 20 m + 39 m + 1 m = 60 m
   c. 20 m + 39 m + 42 m + 18 m + 25 m = 144 m

6.

Distance	Unit
the length of a park	m
from Miami to the North Pole	km
the length of a cell phone	cm
the length of a bus	m & cm

Distance	Unit
around your wrist	cm
the height of a room	m & cm
the length of an airplane trip	km
the length of a grasshopper	cm

Weight in Pounds, p. 71

1. a. 70 pounds   b. 88 pounds   c. 76 pounds
   d. 34 pounds   e. 98 pounds   f. 192 pounds

2-6. Answers will vary.

Weight in Kilograms, p. 73

1. a. 45 kilograms   b. 93 kilograms   c. 69 kilograms

2-7. Answers will vary.

Mixed Review Chapter, p. 75

1.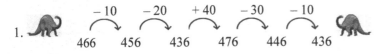
      466   456   436   476   446   436

2. a. △ = 30   b. △ = 9   c. △ = 200

3. a. 806, 816, 826, 836, 846, 856, 866, 876
   b. 542, 532, 522, 512, 502, 492, 482, 472

4. a. 280, 285, 290, 295, 300, 305, 310, 315
   b. 1000, 995, 990, 985, 980, 975, 970, 965

5. a. 300   b. 486   c. 680

6. a.
   b.
   c.
   d.

7. 90 + 90 = 180 km

8.

Distance	Unit
from New York to Chicago	km
the length of your room	m
the length of a pencil	cm

Distance	Unit
around your neck	cm
the width of a butterfly	cm
how far you can throw a ball	m

9. a. 29   b. 480   c. 740

10.

a. 10 past 8 8 : 10	b. 15 till 7 6 : 45	c. 25 past 12 12 : 25	d. half-past 7 7 : 30
e. 9 o'clock 9 : 00	f. 20 till 6 5 : 40	g. 5 till 11 10 : 55	h. 25 till 4 3 : 35

11. Answers will vary.

12. a. triangles   b. hexagons

Puzzle corner: 119, 128, 137, 146, 155, 164, 173, 182, 191

Review Chapter 7, p. 78

1.

Distance	Unit or units
from your house to the grocery store	mi or km
from Miami to New York	mi or km
the distance across the room	m or ft
the height of a bookcase	ft, in, m, or cm

2. about 6 cm *or* about 2 1/2 in.

3. a. ━━━━━━━━━━━━━━━━━

   b. ━━━━━━━━━━━━━━━━━

4. The longer pencil is about 7 cm longer than the shorter one.
   The longer pencil is about 2.5 inches longer than the shorter one.

5. Answers will vary.

Chapter 8: Regrouping in Addition and Subtraction

Adding 3-Digit Numbers in Columns, p. 82

1. a. 308 + 132 = 440   b. 408 + 233 = 641
   c. 226 + 237 = 463   d. 308 + 157 = 465

2. a. 811   b. 763   c. 772   d. 583
   e. 866   f. 989   g. 933   h. 965

3. a. Now there are 675 workers. 547 + 128 = 675
   b. There are 580 balloons left in the bags. 600 − 20 = 580
   c. The total cost is $348. 116 + 116 + 116 = 348

Puzzle Corner:	$\begin{array}{r} 6\ \boxed{3}\ \boxed{1} \\ +\ 1\ \boxed{4}\ 5 \\ \hline 7\ 7\ 6 \end{array}$	$\begin{array}{r} 2\ 2\ \boxed{5} \\ +\ \boxed{3}\ 5\ 5 \\ \hline 5\ 8\ 0 \end{array}$	$\begin{array}{r} 1\ 1\ 6 \\ +\ \boxed{4}\ 5\ \boxed{7} \\ \hline 5\ 7\ 3 \end{array}$	$\begin{array}{r} 2\ 4\ 7 \\ +\ 7\ \boxed{4}\ \boxed{5} \\ \hline 9\ 9\ 2 \end{array}$

Regrouping 10 Tens as a Hundred, p. 84

1. Check the student's work as far as circling the tens.
   a. 130   b. 310
   c. 530   d. 355
   e. 741   f. 656

2. a. 130
   b. 320
   c. 413
   d. 646
   e. 715

3. a. 110, 210, 310   b. 110, 210, 310   c. 140, 340, 540

4. a. 110   b. 310   c. 117   d. 444   e. 815
   f. 139   g. 417   h. 856   i. 729   j. 869

5. a. 110, 170, 210   b. 140, 330, 180   c. 340, 270, 210

6. a. $\begin{array}{r} 1\ 6\ 7 \\ +\ 1\ 9\ 2 \\ \hline 3\ 5\ 9 \end{array}$
   b. $\begin{array}{r} 2\ 4\ 0 \\ +\ 1\ 8\ 2 \\ \hline 4\ 2\ 2 \end{array}$
   c. $\begin{array}{r} 3\ 9\ 1 \\ +\ 4\ 2\ 2 \\ \hline 8\ 1\ 3 \end{array}$
   d. $\begin{array}{r} 6\ 5\ 3 \\ +\ 1\ 9\ 3 \\ \hline 8\ 4\ 6 \end{array}$
   e. $\begin{array}{r} 3\ 7\ 5 \\ +\ 1\ 8\ 4 \\ \hline 5\ 5\ 9 \end{array}$

7. When the elephant and the giraffe ran a race, who won? The elephant, because the giraffe was in the refrigerator.
   (This is a continuation to a silly joke question, which goes: "How do you get a giraffe into a refrigerator?"
   — Open the door, put the giraffe in, close the door.)

Add in Columns: Regrouping Twice, p. 88

1. a. 58 + 83 = 141   b. 86 + 79 = 165
   c. 155 + 69 = 224   d. 177 + 136 = 313

2. a. 241   b. 342   c. 422   d. 652

3. Mary circled 10 tens to make another hundred, and she circled the ones to form a 10.
   Then she added the tens with the new group, which made 40, and she added the hundreds,
   which made 600. This gave her a total of 640.

Add in Columns: Regrouping Twice, cont.

4. a. 767   b. 524   c. 810   d. 651   e. 860   f. 601
   g. 624   h. 819   i. 759   j. 931   k. 832   l. 930

5. *First I check the hundreds: 600 + 300 makes <u>900</u>.*
    Then I add the <u>tens</u>, and I get 50 + <u>30</u> = <u>80</u>.
    Lastly in the ones, there's just 0 and 1, which is 1.
    Okay, so I have these parts: 900, 80, and <u>1</u>, so that makes <u>981</u>.

6. a. 148 + 148 = 296 miles.
   b. $128 + $129 + $549 = $896 was the total cost.
   c. 156 + 156 + 124 = 436 feet.

Puzzle corner:

```
   2  3  2        2  8  1        1  6  9        1  8  8
+  1  5  9     +  2  3  6     +  5  5  5     +  7  1  2
---------      ---------      ---------      ---------
   3  9  1        5  1  7        7  2  4        9  0  0
```

Regrouping in Subtraction, Part 1, p. 92

1.

a. 3 tens 0 ones ⟹ 2 tens 10 ones	b. 3 tens 6 ones ⟹ 2 tens 16 ones
c. 5 tens 1 one ⟹ 4 tens 11 ones	d. 5 tens 5 ones ⟹ 4 tens 15 ones
e. 4 tens 0 ones ⟹ 3 tens 10 ones	f. 3 tens 7 ones ⟹ 2 tens 17 ones

2. a. What is left? <u>2</u> tens <u>8</u> ones
   b. 5 tens 4 ones → 4 tens 14 ones. What is left? <u>2</u> tens <u>7</u> ones.
   c. 4 tens 3 ones → 3 tens 13 ones. What is left? <u>1</u> ten <u>8</u> ones.
   d. 6 tens 1 one → 5 tens 11 ones. What is left? <u>1</u> ten <u>7</u> ones.

3.

3. a. 5 tens 5 ones ⟹			b. 7 tens 2 ones ⟹		
	4 tens	15 ones		6 tens	12 ones
−	3 tens	7 ones	−	3 tens	5 ones
	1 ten	8 ones		3 tens	7 ones
c. 6 tens 0 ones ⟹			d. 6 tens 4 ones ⟹		
	5 tens	10 ones		5 tens	14 ones
−	2 tens	7 ones	−	3 tens	8 ones
	3 tens	3 ones		2 tens	6 ones
e. 7 tens 6 ones ⟹			f. 5 tens 0 ones ⟹		
	6 tens	16 ones		4 tens	10 ones
−	4 tens	7 ones	−	2 tens	2 ones
	2 tens	9 ones		2 tens	8 ones
g. 8 tens 1 one ⟹			h. 6 tens 3 ones ⟹		
	7 tens	11 ones		5 tens	13 ones
−	6 tens	5 ones	−	2 tens	8 ones
	1 ten	6 ones		3 tens	5 ones

4. a. Jessica now has 13 pencils. 27 − 10 − 4 = 13.
   b. Three more pencils. 13 − 10 = 3.
   c. Nine more pencils. 13 − 4 = 9.

1.

a. 6 tens 0 ones → 5 tens 10 ones Take away 3 tens, 9 ones. $\begin{array}{c c}\;6 & 0\\ -\;3 & 9 \\ \hline \;2 & 1\end{array}$	b. 7 tens 1 one → 6 tens 11 ones Take away 1 ten, 6 ones. $\begin{array}{c c}\;7 & 1\\ -\;1 & 6 \\ \hline \;5 & 5\end{array}$
c. 3 tens, 5 ones → 2 tens 15 ones Take away 1 ten, 7 ones. $\begin{array}{c c}\;3 & 5\\ -\;1 & 7 \\ \hline \;1 & 8\end{array}$	d. 8 tens → 7 tens 10 ones Take away 3 tens, 4 ones. $\begin{array}{c c}\;8 & 0\\ -\;3 & 4 \\ \hline \;4 & 6\end{array}$
e. 7 tens, 6 ones → 6 tens 16 ones Take away 4 tens, 8 ones. $\begin{array}{c c}\;7 & 6\\ -\;4 & 8 \\ \hline \;2 & 8\end{array}$	f. 9 tens → 8 tens 10 ones Take away 5 tens, 1 one. $\begin{array}{c c}\;9 & 0\\ -\;5 & 1 \\ \hline \;3 & 9\end{array}$
g. 5 tens, 4 ones → 4 tens 14 ones Take away 2 tens, 5 ones. $\begin{array}{c c}\;5 & 4\\ -\;2 & 5 \\ \hline \;2 & 9\end{array}$	h. 8 tens → 7 tens 10 ones Take away 4 tens, 7 ones. $\begin{array}{c c}\;8 & 0\\ -\;4 & 7 \\ \hline \;3 & 3\end{array}$
i. 7 tens, 4 ones → 6 tens 14 ones Take away 3 tens, 8 ones. $\begin{array}{c c}\;7 & 4\\ -\;3 & 8 \\ \hline \;3 & 6\end{array}$	j. 4 tens 7 ones → 3 tens 17 ones Take away 2 tens, 9 ones. $\begin{array}{c c}\;4 & 7\\ -\;2 & 9 \\ \hline \;1 & 8\end{array}$

2. In this exercise, the child is asked to check each subtraction with addition. Adding back is just one method of checking. Estimating the result beforehand is another method of checking - it won't tell you if you got it exactly right but it does check that the result is at least reasonable. Yet another way is to solve the same problem with some different method (there might not always be a "different method"). It is not good that children get used to always checking answers for math problems from the back of the book or from their teacher. They should get used to checking the results themselves. You can encourage the students to check their results even if there is no written instruction to do so.

   a. 29   b. 62   c. 27   d. 55   e. 57   f. 26   g. 38   h. 39   i. 22   j. 38   k. 13   l. 38

Puzzle corner:

$\begin{array}{c c}\;9 & 3\\ -\;1 & 8 \\ \hline \;7 & 5\end{array}$ $\begin{array}{c c}\;8 & 3\\ -\;6 & 7 \\ \hline \;1 & 6\end{array}$ $\begin{array}{c c}\;8 & 0\\ -\;3 & 8 \\ \hline \;4 & 2\end{array}$ $\begin{array}{c c}\;8 & 2\\ -\;1 & 4 \\ \hline \;6 & 8\end{array}$ $\begin{array}{c c}\;6 & 2\\ -\; & 9 \\ \hline \;5 & 3\end{array}$

Regrouping in Subtraction, Part 3, p. 98

1. a. no   b. yes   c. yes

2. a. YES; 44   b. NO; 25   c. YES; 26   d. YES; 52   e. NO; 72   f. YES; 24   g. NO; 10   h. YES; 19

3. a. 44, 48   b. 6, 5   c. 92, 4

4. a. $45 - 27 = 18$ b. $34 - 13 = 21$ c. $31 - 9 - 8 = 14$

5. a. $22 - 9 = 13$

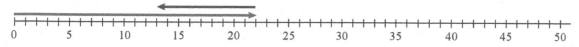

   b. $36 - 12 = 24$

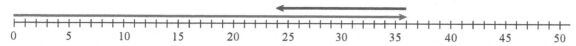

   c. $44 - 17 = 27$

6. a. 94   b. 57, 82   c. 12, 61   d. 42, 99   e. 37, 60   f. 18, 66   g. 13, 54   h. 46, 85

7. a. Emily picked 16 rows more.
   b. Peter sold 27 more tickets.
   c. They sold 97 tickets.
   d. The doll with the pink dress costs $40.
   e. The total cost was $85.

Word Problems, p. 102

1. a. Andy has three more goats than Isabella.   b. Sandy has 23 goats.
   c. Henry has ten goats.   d. Daniel has 43 cows.

2. a. Mason has 22 tennis balls.
   b. Logan has 34 tennis balls.
   c. The yellow dress costs $25.
   d. Together, the blue and yellow dresses would cost $66.

3. a. 2   b. 22   c. 8   d. 7   e. 64   f. 95   g. 51   h. 11   j. 42

4. a. $11 - 8 = 3 + 3 = 6$ km   Zoe rode 6 km less than Ryan for 2 days.
   b. $32 - 8 = 24 + 32 = 56$ Together, the two girls have 56 dolls.

Puzzle corner: a. ▲ = 8   b. ▲ = 26   c. △ = 50   d.-f. Answers will vary.

Mental Subtraction, Part 1, p. 105

1. a. $(51 - \underline{1}) - 4 = 46$ b. $(62 - \underline{2}) - 5 = 55$ c. $(33 - \underline{3}) - 1 = 29$ d. $(92 - \underline{2}) - 3 = 87$
   e. $(75 - \underline{5}) - 1 = 69$ f. $(63 - \underline{3}) - 4 = 56$ g. $(35 - \underline{5}) - 2 = 28$ h. $(74 - \underline{4}) - 1 = 69$

2. a. 44, 46, 48, 45   b. 38, 37, 39, 36

3. a. 5, 15, 35   b. 9, 19, 29   c. 3, 43, 23   d. 6, 56, 36   e. 5, 25, 85   f. 8, 78, 28

4. a. Amy has $25 now.
   b. Peter needs $5 more.
   c. 47 are red.

5. a. 29, 66   b. 56, 28   c. 43, 87   d. 55, 75

Mental Subtraction, Part 2, p. 107

1. a. 8   b. 2   c. 7   d. 4   e. 8   f. 7
   g. 7   h. 8   i. 7   j. 5   k. 9   l. 7

2. a. Larry has 53 cars. 46 + 7 = 53. Mickey has 39 cars. 46 − 7 = 39.
   b. Andy has $10 now. After he bought the gift, he had $33 − $28 = $5. Then he got $5 more so he had $5 + $5 = $10.

3. a. 56   b. 25   c. 21   d. 23   e. 16   f. 32   g. 33   h. 13   i. 11

4. a. There were 8 more books. 42 − 34 = 8.
   b. They both now have 38 books. The second shelf has now 42 − 4 = 38, and the first shelf has 34 + 4 = 38.

5. 35 − 18 = 17

6. a. 14   b. 61   c. 63   d. 26   e. 34   f. 62

7. a. Terry has 43 pages left to read. 95 − 52 = 43.
   b. Now she has 34 pages left to read. You can solve this in many ways. For example: she has now read 52 + 9 = 61 pages.
      So, she has 95 − 61 = 34 pages left. Or, since she had 43 pages left to read earlier, now she has 43 − 9 = 34 pages left.

8. There are various methods you can use, such as breaking a ten into ones to subtract; or adding up
   to the next number; or subtracting in parts. If the student cannot answer this question, then he/she
   needs to review what has already been taught in previous lessons.

   a. 16   b. 34

Puzzle corner:

a. △ = 12   b. △ = 9   □ = 13   c. △ = 4   □ = 18

Regrouping One Ten as Ten Ones with 3-Digit Numbers, p. 110

1.

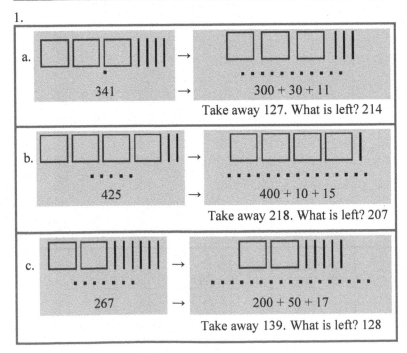

2. a. 815   b. 467   c. 417

3. a. 618; 618 + 205 = 823   b. 216; 216 + 455 = 671
   c. 305; 305 + 115 = 420   d. 308; 308 + 208 = 516
   e. 106; 106 + 627 = 733   f. 378; 378 + 309 = 687

Regrouping One Ten as Ten Ones with 3-Digit Numbers, cont.

4.

a.	b.
120 + 120 = 240	1 + 119 = 120
121 + 119 = 240	2 + 118 = 120
122 + 118 = 240	3 + 117 = 120
123 + 117 = 240	4 + 116 = 120
124 + 116 = 240	5 + 115 = 120
125 + 115 = 240	6 + 114 = 120
126 + 114 = 240	7 + 113 = 120
127 + 113 = 240	8 + 112 = 120
Till you reach 240 + 0 = 240.	Till you reach 120 + 0 = 120.

5. a. There are 730 days in two years. 365 + 365 = 730
 b. They still have 104 km to travel. 173 − 69 = 104.
 c. He has 283 pages left to read. 581 − 235 = 346

Regrouping One Hundred as 10 Tens, p. 113

1. a. 160 b. 160
 c. 151 d. 80

2. a. 175
 b. 360
 c. 70
 d. 84
 e. 272

3.

It would be easy, except I cannot subtract <u>8</u> tens from <u>4</u> tens. So, I need to take one of the <u>9</u> hundreds and break it down as tens. So, now I will have only <u>8</u> hundreds but I will now get <u>14</u> tens. Now I can subtract.

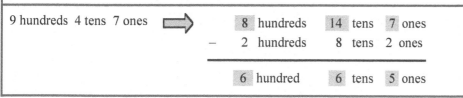

9 hundreds 4 tens 7 ones	⟹	8 hundreds	14 tens	7 ones
	−	2 hundreds	8 tens	2 ones
		6 hundred	6 tens	5 ones

4. a. 184 b. 172

5. a. 780 b. 434 c. 92 d. 237
 e. 444 f. 552 g. 162 h. 372

6. a. The second book has 210 pages. 270 − 60 = 210
 b. Liz got 241 more points than Hannah. 433 − 192 = 241
 c. Hannah got 308 points. 215 + 93 = 308
 d. Micah caught 25 worms. 14 + 11 = 25. Together they caught 39 worms. 25 + 14 = 39

Puzzle corner:

4	4	5		6	8	4		9	8	1		9	6	0
− 1	5	3		− 3	5	8		− 2	5	5		− 4	5	5
2	9	2		3	2	6		7	2	6		5	0	5

Graphs and Problems, p. 117

1.

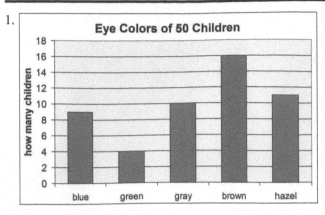

a. 27 children   b. 12 children   c. 41 children

2. a. No, together they have one less marble than Jim.
   b. You have read 58 pages. You have 37 pages left to read.

3. a. 90 birds   b. 25 Australian animals
   c. 85 together   d. 35 more   e. 135 animals

4. a. 22   b. 61   c. 32

Euclid's Game, p. 119

1. No.
2. All the numbers that get marked are also even.
3. All the numbers that get marked will also be multiples of 5.
4. a. No. You cannot mark numbers that are greater than 90.
   b. Yes.
   c. No. You will only get to mark multiples of 10.
   d. Yes.

Mixed Review Chapter 8, p. 122

1.

8	9	10	11	12	13	14	15	16	17
16	18	20	22	24	26	28	30	32	34

2. a. Each girl got 15 marbles.
   b. There are 14 kilograms of potatoes left.
   c. Katy now has $21. Half of $60 is $30. Then, $30 − $9 = $21.
   d. Mom had 16 apples to start with.

3. a. 139   b. 471   c. 204   d. 336   e. 761   f. 950

4. a. These months have 31 days - January, March, May, July, August, October and December.
   b. These months have 30 days - April, June, September, November
   c. February has 28 days, except for every four years in a leap year, when it has 29 days.

5.

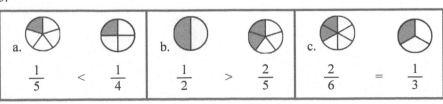

Mixed Review Chapter 8, cont.

6. a. ▲ = 80 b. ▲ = 90 c. ▲ = 160
    a. Subtract 800 and 720. That works because subtraction is the opposite operation of addition.
    b. Subtract 200 and 110. The reason being, when the subtrahend is missing, it is like one "part" is missing.
      To find that part, subtract the other part from the total.
    c. Add 70 and 90. The minuend is missing, which is the "total" in the subtraction, so to find it, add the "parts."

7.

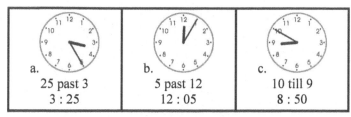

a.	b.	c.
25 past 3	5 past 12	10 till 9
3 : 25	12 : 05	8 : 50

8. Check the students' work.

Review Chapter 8, p. 124

1. a. 692   b. 417   c. 718   d. 764

2. The three bicycles cost a total of $462. Add: 154 + 154 + 154 = 462.

3. a. 120, 820   b. 180, 290   c. 740, 550

4. It is 380 feet all of the way around. Add 120 + 70 + 120 + 70 = 380.

5. a. 34, 88   b. 15, 63   c. 35, 84   d. 723, 882   e. 165, 556   f. 304, 550

6. a. 8, 48   b. 8, 88   c. 5, 4   d. 9, 5   e. 16, 13   f. 43, 29

7. a.   2 **5** 4   b.   5 **8** 9   c.   2 0 **6**   d.   6 8 **1**
       + 4 7 7   + **3** 2 5   + 6 **8** 6   + **2** 1 9
       ———   ———   ———   ———
         7 3 1   9 1 4   8 9 2   9 0 0

8. a. 39 people. 52 − 13 = 39
   b. 39 toys. 23 + 16 = 39
   c. 17 toys. 33 − 16 = 17
   d. 700 points. 465 + 145 + 90 = 700
   e. 40 jumping jacks. 26 + 14 = 40.

9. a.

CHILD	POINTS
Charlie	15
Bill	24
Amy	28
Cindy	21
Sarah	19

   b. 4 points   c. 6 points

Puzzle corner:

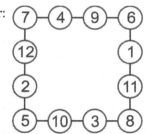

Chapter 9: Money

Counting Coins Review, p. 130

1. a. 40¢   b. 35¢
   c. 38¢   d. 39¢
   e. 45¢   f. 37¢
   g. 63¢   h. 34¢

2. Answers may vary.

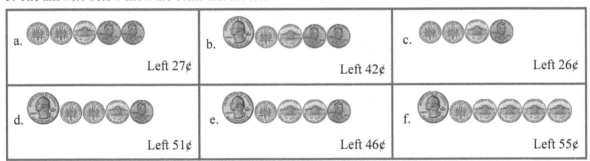

3. a. 80¢   b. 70¢   c. 37¢   d. 36¢   e. 92¢   f. 96¢   g. 92¢   h. 103¢

4.

a. 26¢	b. 40¢
c. 52¢	d. 77¢

5. The answers below show the coins that are left:

a. Left 27¢	b. Left 42¢	c. Left 26¢
d. Left 51¢	e. Left 46¢	f. Left 55¢

6. a. 8¢   b. 45¢   c. 60¢   d. 85¢   e. 96¢   f. 41¢

7.

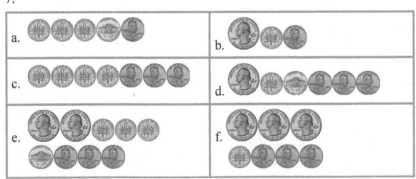

8.

a.	b.	c.
$92¢ + 8¢ = 100¢$	$70¢ + 30¢ = \$1$	$40¢ + 60¢ = \$1$
$80¢ + 20¢ = \$1$	$74¢ + 26¢ = \$1$	$33¢ + 67¢ = \$1$
$79¢ + 21¢ = \$1$	$64¢ + 36¢ = \$1$	$45¢ + 55¢ = \$1$
$50¢ + 50¢ = \$1$	$58¢ + 42¢ = \$1$	$31¢ + 69¢ = \$1$

Change, p. 134

1. a. give 25¢, change 5¢   b. give 50¢, change 20¢
   c. give 40¢, change 5¢   d. give 20¢, change 3¢
   e. give 30¢, change 8¢   f. give 15¢, change 4¢
   g. give 75¢, change 15¢   h. give 100¢, change 20¢

2. a. Use 2 quarters, 1 dime to pay; change is 5¢
   b. Use one quarter, one dime to pay; change is 2¢
   c. Use one quarter, two dimes, one nickel to pay; change is 4¢
   d. Use 2 quarters, four dimes; change 2¢
   e. Use two quarters, three dimes; change 5¢
   f. Use one quarter, three dimes, two nickels; change 2¢

3. a. 30¢ b. 9¢ c. 6¢
   d. 30¢ e. 30¢ f. 13¢

4. b. 35¢ total, 15¢ change
   c. 16¢ total, 4¢ change
   d. 20¢ total, 5¢ change
   e. 45¢ total, 5¢ change

Dollars, p. 137

1. a. $1.20 b. $5.16
   c. $5.40 d. $1.26
   e. $1.50 f. $5.56
   g. $1.66 h. $1.84

2. b. $7.21
   c. $2.43 d. $2.60
   e. $7.88 f. $9.71

3. a. $0.30 b. $0.02 c. $0.07
   d. $0.60 e. $0.10 f. $0.21

4. a. $1.30 b. $1.11
   c. $1.41 d. $1.07

5. Answers may vary since there are more than one way to use the coins for the same cent-value.
   a. One dollar, three dimes, and two pennies.   b. Two dollars, one nickel, and one penny.
   c. Two dollars, two quarters, and four pennies.   d. Three dollars, three quarters, and one nickel.

Counting Change, p. 140

1. a. Two dimes and two pennies; 22¢
   b. One quarter and one dime; 35¢
   c. Two quarters and three pennies; 53¢
   d. One quarter, two dimes, and three pennies; 48¢

2. a. Three quarters and one dime; 85¢
   b. Two dimes; 20¢
   c. Two dimes and two pennies; 22¢
   d. Two quarters, one dime, one nickel and three pennies; 68¢

3. a. $0.06 b. $0.12
   c. $0.03 d. $0.20
   e. $0.75 f. $0.15

Adding Money Amounts, p. 142

1. a. $0.85   b. $1.50   c. $4.81

2. a. $2.17   b. $1.20

3. a. $3.02   b. $3.80
   c. $4.77   d. Answers vary.

4. a. change $1.98
   b. change $0.20
   c. $0.23

Mixed Review Chapter 9, p. 144.

1. a. 316   b. 433   c. 643

2. a. Together they have 143 marbles.
   b. Mom has $80 left.
   c. He has 72 pages left to read.

3. a. 65, 93   b. 283

4. The answer lists the base side of the triangle first.

Triangle	in inches	in centimeters
Side 1	2 in.	5 cm
Side 2	3 in.	8 cm
Side 3	3 in.	8 cm

5. a. Chris   b. 5 more   c. 4 fewer fish   d. 15 together

6.

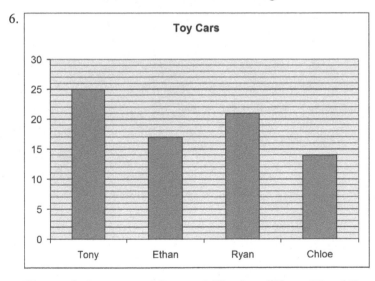

a. 17 cars   b. 8 cars   c. 35 cars   d. Yes, he will have 26 and Tony 25.

Puzzle corner:

1. a. 33¢   b. 50¢

2. a. One quarter and three pennies.
   b. Three quarters, one dime, one nickel, and three pennies.

3. a. $2.35   b. $7.19
   c. $0.45   d. $0.49

4. a. You give 70¢, your change is 5¢   b. You give $1.00, your change is 8¢

5.

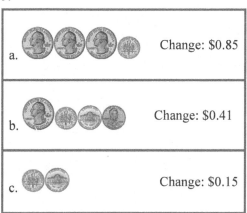

a.	Change: $0.85
b.	Change: $0.41
c.	Change: $0.15

6. Alex has 77¢. Together they have $2.03.

7. a. $1.76   b. $3.61

Chapter 10: Exploring Multiplication

Many Times the Same Group, p. 151

1. Check the student's work.

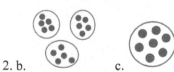

2. b. 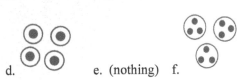 c.

d.   e. (nothing)   f.

g. (nothing)   h.   i.

3. b. 3 times 2   c. 5 times 1   d. 1 times 9   e. 3 times 4   f. 3 times 0

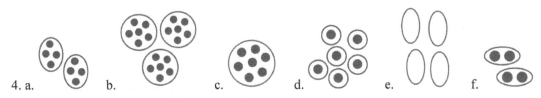

4. a.   b.   c.   d.   e.   f.

5. b. $3 \times 6 = 18$ c. $4 \times 0 = 0$ d. $3 \times 1 = 3$ e. $1 \times 3 = 3$ f. $3 \times 4 = 12$

6. a. 8   b. 10   c. 4   d. 10   e. 16   f. 9

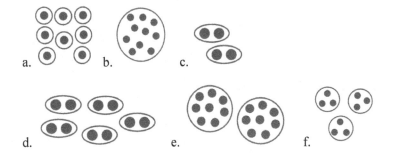

a.   b.   c.

d.   e.   f.

Multiplication and Addition, p. 154

1.

a. $2 + 2 + 2 = 6$ $3 \times 2 = 6$		b. $3 + 3 + 3 = 9$ $3 \times 3 = 9$	
c. $\begin{array}{r} 2 \\ 2 \\ 2 \\ + 2 \\ \hline 8 \end{array}$ $4 \times 2 = 8$	d. $\begin{array}{r} 6 \\ + 6 \\ \hline 12 \end{array}$ $2 \times 6 = 12$	e. $\begin{array}{r} 4 \\ 4 \\ + 4 \\ \hline 12 \end{array}$ $3 \times 4 = 12$	

2.

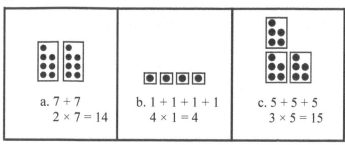

a. 7 + 7 2 × 7 = 14	b. 1 + 1 + 1 + 1 4 × 1 = 4	c. 5 + 5 + 5 3 × 5 = 15

3.

a. _4_ groups, _2_ scissors in each. _4_ × _2_ scissors = 8 scissors _2_ + _2_ + 2 + 2	b. 4 groups, 3 rams in each. 4 × 3 rams = 12 rams 3 + 3 + 3 + 3
c. 3 groups, 2 bears in each. 3 × 2 bears = 6 bears 2 + 2 + 2	d. 2 groups, 4 carrots in each. 2 × 4 carrots = 8 carrots 4 + 4
e. 3 groups, 5 books in each. 3 × 5 books = 15 books 5 + 5 + 5	f. 4 groups, 5 bulbs in each. 4 × 5 bulbs = 20 bulbs 5 + 5 + 5 + 5
g. 5 groups, 3 scissors in each. 5 × 3 scissors = 15 scissors 3 + 3 + 3 + 3 + 3	h. 2 groups, 7 rams in each. 2 × 7 rams = 14 rams 7 + 7

4.

a. 2 + 2 + 2 + 2 + 2 = 10 5 × 2 = 10	b. 1 + 1 + 1 = 3 3 × 1 = 3
c. 3 + 3 + 3 = 9 3 × 3 = 9	d. 2 + 2 = 4 2 × 2 − 4
e. 2 + 2 + 2 + 2 + 2 + 2 + 2 = 14 7 × 2 = 14	f. 6 + 6 = 12 2 × 6 = 12
g. 3 + 3 + 3 + 3 + 3 + 3 + 3 + 3 = 24 8 × 3 = 24	h. 1 + 1 + 1 + 1 = 4 4 × 1 = 4

5.

a.	b.	c.	d.
3 × 7 = 21	2 × 8 = 16	4 × 4 = 16	5 × 2 = 10
e.	f.	g.	h.
4 × 2 = 8	10 × 1 = 10	5 × 3 = 15	7 × 2 = 14

Multiplying on a Number Line, p. 158

1. a. $7 \times 2 = 14$ b. $4 \times 4 = 16$ c. $3 \times 3 = 9$ d. $7 \times 1 = 7$

2.

3. a. 15, 12   b. 24, 21   c. 18, 9   d. 6, 27

4. a. 8, 3   b. 6, 5   c. 7, 4   d. 2, 1

5.
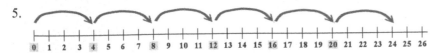

6. a. 8, 16   b. 24, 28   c. 32, 12   d. 20, 4

7. a. 6, 2   b. 0, 3   c. 4, 6   d. 5, 1

8. a.

$6 \times 4 = 24$

b.

$5 \times 5 = 25$

c.

$6 \times 1 = 6$

d.
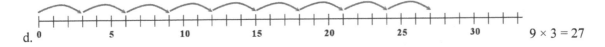
$9 \times 3 = 27$

e.

$3 \times 10 = 30$

9. a. 12   b. 22   c. 20   d. 18

Puzzle corner: a. 2   b. 3   c. 6   d. 3   e. 7

Multiplication Practice, p. 161

1. a. 6, 8, 10   b. 6, 9, 12   c. 30, 50, 60

2. a. $2 \times 11 = 22$ b. $4 \times 0 = 0$ c. $5 \times 10 = 50$
   d. $5 \times 1 = 5$ e. $3 \times 4 = 12$ f. $3 \times 20 = 60$
   g. $3 \times 300 = 900$ h. $7 \times 10 = 70$

3.
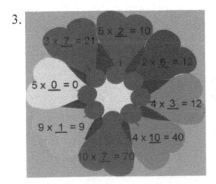

Multiplication Practice, cont.

4. 0, 2, 4, 6, 8, 10, 12, 14, 16, 18, 20, 22, 24

$1 \times 2 = 2$	$4 \times 2 = 8$	$7 \times 2 = 14$	$10 \times 2 = 20$
$2 \times 2 = 4$	$5 \times 2 = 10$	$8 \times 2 = 16$	$11 \times 2 = 22$
$3 \times 2 = 6$	$6 \times 2 = 12$	$9 \times 2 = 18$	$12 \times 2 = 24$

5. 0, 5, 10, 15, 20, 25, 30, 35, 40, 45, 50, 55, 60

$1 \times 5 = 5$	$4 \times 5 = 20$	$7 \times 5 = 35$	$10 \times 5 = 50$
$2 \times 5 = 10$	$5 \times 5 = 25$	$8 \times 5 = 40$	$11 \times 5 = 55$
$3 \times 5 = 15$	$6 \times 5 = 30$	$9 \times 5 = 45$	$12 \times 5 = 60$

6. 0, 10, 20, 30, 40, 50, 60, 70, 80, 90, 100, 110, 120

$1 \times 10 = 10$	$4 \times 10 = 40$	$7 \times 10 = 70$	$10 \times 10 = 100$
$2 \times 10 = 20$	$5 \times 10 = 50$	$8 \times 10 = 80$	$11 \times 10 = 110$
$3 \times 10 = 30$	$6 \times 10 = 60$	$9 \times 10 = 90$	$12 \times 10 = 120$

7.

a. $2 \times 5 = 10$ $3 \times 3 = 9$	b. $10 \times 10 = 100$ $9 \times 2 = 18$	c. $5 \times 9 = 45$ $3 \times 2 = 6$	d. $5 \times 5 = 25$ $3 \times 7 = 21$
e. $7 \times 2 = 14$ $5 \times 4 = 20$	f. $7 \times 5 = 35$ $3 \times 4 = 12$	g. $10 \times 6 = 60$ $2 \times 8 = 16$	h. $12 \times 10 = 120$ $2 \times 12 = 24$

Mixed Review Chapter 10, p. 163

1. a. 624, 244   b. 707, 277   c. 72, 4

2. a. $1.45   b. $5.21

3. Brett has 100¢. Together they have 157¢.

4. a. The total cost is $1.10 and the change is $3.90.
   b. The total cost is $1.60 and the change is $0.40.

5.

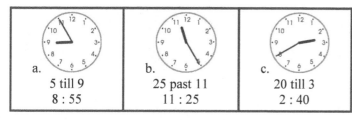

a. 5 till 9 8 : 55	b. 25 past 11 11 : 25	c. 20 till 3 2 : 40

6. a. 9:00   b. 11:30   c. 2:45

7. a. 425, 652   b. 363, 548

8. 350 meters all the way around.

9. a. ▲ = 40   b. ▲ = 90   c. ▲ = 120

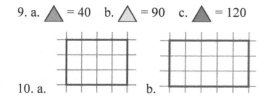

10. a.   b.

11. a. 6 faces, squares   b. 6 faces, rectangles

12. a. cone   b. pyramid   c. cylinder   d. cube

1. a. $5 \times 1 = 5$

   b. $2 \times 10 = 20$

   c. $3 \times 2 = 6$

2. a. 18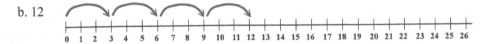

   b. 12

3. a. $3 + 3 + 3 = 9$ b. $2 + 2 + 2 + 2 = 8$

4. a. 6, 20, 40   b. 9, 60, 7   c. 3, 0, 20

5. a. 12   b. 8   c. 8   d. 4   e. 12   f. 4

6. $5 + 5 + 5 + 5 = 20$; $5 + 5 + 5 = 15$; $5 + 5 = 10$; $1 \times 5 = 5$; $2 \times 5 = 10$; $3 \times 5 = 15$; $4 \times 5 = 20$;
   $20 - 5 = 15$; $20 - 5 - 5 = 10$; $20 - 5 - 5 - 5 = 5$; $20 - 5 - 5 - 5 - 5 = 0$;

7. a. Fifty flowers. $5 \times 10 = 50$
   b. Jim has more. Now he has ten more.

8.

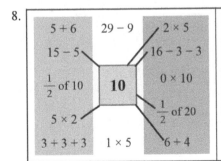 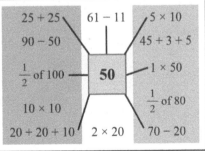

Test Answer Keys

Math Mammoth Grade 2 Tests Answer Key

Chapter 1 Test

Grading

My suggestion for grading the chapter 1 test is below. The total is 24 points. Divide the student's score by the total of 24 to get a decimal number, and change that decimal to percent to get the student's percentage score.

Question	Max. points	Student score
1	8 points	
2	2 points	
3	6 points	

Question	Max. points	Student score
4	2 points	
5	6 points	
Total	24 points	

1. a. 60, 52   b. 37, 90   c. 56, 35   d. 85, 28

2. a.   b.

3.

a. $2 + 8 = 10$	b. $6 + 3 = 9$	c. $7 + 8 = 15$
$8 + 2 = 10$	$3 + 6 = 9$	$8 + 7 = 15$
$10 - 8 = 2$	$9 - 6 = 3$	$15 - 7 = 8$
$10 - 2 = 8$	$9 - 3 = 6$	$15 - 8 = 7$

4. She read 46 pages.

5.

Number	Even?	Odd?
4	X	
10	X	

Number	Even?	Odd?
9		X
16	X	

Number	Even?	Odd?
11		X
18	X	

Chapter 2 Test

Grading

My suggestion for grading the chapter 2 test is below. The total is 29 points. Divide the student's score by the total of 29 to get a decimal number, and change that decimal to percent to get the student's percentage score.

Question	Max. points	Student score
1	16 points	
2	4 points	
3	4 points	

Question	Max. points	Student score
4	5 points	
Total	29 points	

1.

a. 2:05 5 past 2	b. 5:45 15 till 6	c. 2:40 20 till 3	d. 6:30 half past 6
e. 3:35 25 till 4	f. 7:25 25 past 7	g. 8:50 10 till 9	h. 4:00 4 o'clock

2.

Time now	3:50	7:25
5 minutes later	3:55	7:30

Time now	9 AM	12 noon
1 hour later	10 AM	1 PM

3.

a. 20 past 4 4:20	b. 15 past 11 11:15	c. 15 till 12 11:45	d. 25 till 7 6:35

4.

from	5 AM	8 AM	2 AM	10 AM	11 AM
to	12 noon	2 PM	3 PM	10 PM	6 PM
hours	7	6	13	12	7

Chapter 3 Test

Grading

My suggestion for grading the chapter 3 test is below. The total is 31 points. Divide the student's score by the total of 31 to get a decimal number, and change that decimal to percent to get the student's percentage score.

Question	Max. points	Student score
1	12 points	
2	6 points	
3	3 points	

Question	Max. points	Student score
4	6 points	
5	4 points	
Total	31 points	

1. a. 15, 13   b. 17, 12   c. 7, 9   d. 3, 9   e. 13, 13   f. 14, 15

2. a. 9, 9 + 5 = 14   b. 3, 3 + 8 = 11   c. 9, 9 + 8 = 17

3. a. =   b. >   c. >

4. a. 5, 8   b. 8, 6   c. 7, 7

5. a. 16 teddy bears. 7 + 9 = 16
   b. I still need to save $2. $7 + $5 = $12. And $12 + $2 = $14.

Chapter 4 Test

Grading

My suggestion for grading the chapter 4 test is below. The total is 21 points. Divide the student's score by the total of 21 to get a decimal number, and change that decimal to percent to get the student's percentage score.

Question	Max. points	Student score
1	5 points	
2	6 points	
3	4 points	

Question	Max. points	Student score
4	6 points	
Total	21 points	

1. a. 85   b. 122   c. 87   d. 61   e. 104

2. a. 59, 23   b. 73, 35   c. 93, 51

3. a. 23, 24   b. 59, 76

4. a. She worked 15 more hours. 28 − 13 = 15
   b. The total cost is $46. 12 + 17 + 17 = 46
   c. There are now 22 birds in the tree. 15 + 9 − 2 = 22

Chapter 5 Test

Grading

My suggestion for grading the chapter 5 test is below. The total is 22 points. Divide the student's score by the total of 22 to get a decimal number, and change that decimal to percent to get the student's percentage score.

Question	Max. points	Student score
1	4 points	
2	2 points	
3	2 points	
4	4 points	

Question	Max. points	Student score
5	4 points	
6	6 points	
Total	22 points	

1.

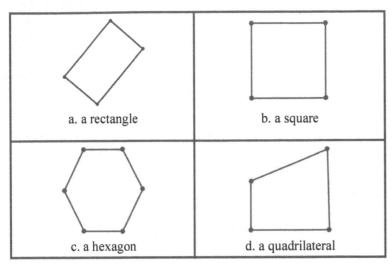

2.

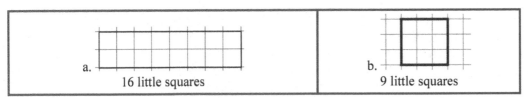

3. Answers will vary. Please check the student's work.

4. a. 1/5   b. 5/6   c. 3/3   d. 2/4

5.

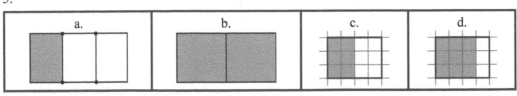

6.

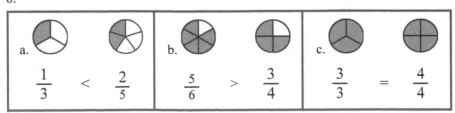

Chapter 6 Test

Grading

My suggestion for grading the chapter 6 test is below. The total is 33 points. Divide the student's score by the total of 33 to get a decimal number, and change that decimal to percent to get the student's percentage score.

Question	Max. points	Student score
1	5 points	
2	2 points	
3	4 points	
4	4 points	
5	2 points	

Question	Max. points	Student score
6	4 points	
7	3 points	
8	6 points	
9	3 points	
Total	33 points	

1. a. 475, 480, 485, 490, 495, 500, 505
   b. 376, 386, 396, 406, 416, 426, 436

2. a. $235 = 200 + 30 + 5$ b. $805 = 800 + 0 + 5$

3. a. 688, 460   b. 285, 106

4. a. $<$ b. $>$ c. $<$ d. $>$

5. a. $689 < 869 < 986$ b. $245 < 452 < 524$

6. a. $256 = 256$ b. $809 < 890$ c. $462 > 246$ d. $703 < 706$

7. a. ▲ $= 40$ b. △ $= 8$ c. ▲ $= 500$

8. a. 565, 248   b. 402, 380   c. 278, 94

9. a. Natalie counted 25 cars
   b. Jayden counted 35 cars.
   c. Natalie counted five more cars than Caleb.

Chapter 7 Test

Grading

My suggestion for grading the chapter 7 test is below. The total is 17 points. Divide the student's score by the total of 17 to get a decimal number, and change that decimal to percent to get the student's percentage score.

Question	Max. points	Student score
1	4 points	
2	4 points	
3	2 points	
4	1 points	

Question	Max. points	Student score
5	1 points	
6	4 points	
7	1 points	
Total	17 points	

1. (a), (c) and (d) do not make sense.

2. #1 is about 4 inches, 10 cm   # 2 is about 5 inches, 13 cm

3. a. ▬▬▬▬▬▬▬▬▬▬▬▬▬▬▬▬

   b. ▬▬▬▬▬▬▬▬▬▬▬▬▬

4. centimeter, inch, foot, kilometer

5. 3 meters

6.

Distance	Unit
from Florida to California	km
around your head	cm

Distance	Unit
length of a garden	m
height of a room	m

7. Answers will vary.

Chapter 8 Test

Grading

My suggestion for grading the chapter 8 test is below. The total is 27 points. Divide the student's score by the total of 27 to get a decimal number, and change that decimal to percent to get the student's percentage score.

Question	Max. points	Student score
1	4 points	
2	4 points	
3	6 points	

Question	Max. points	Student score
4	2 points	
5	11 points	
Total	27 points	

1. a. 654   b. 937   c. 782   d. 35

2. a. 737, 964   b. 376, 748

3. a. 120, 330   b. 750, 54   c. 41, 2

4. Nancy's total bill was $387.

5. a. There are 182 sacks of wheat left. $250 - 68 = 182$
   b. There are 19 black kittens. $52 - 15 - 18 = 19$
   c. The three bags weigh 80 pounds. $15 + 15 + 50 = 80$
   d. They sold 28 coffee makers in February. $47 - 19 = 28$
      They sold 75 coffee makers in two months. $47 + 28 = 75$
   e. Grandpa walked a total of 720 meters. $300 + 300 + 120 = 720$

Chapter 9 Test

Grading

My suggestion for grading the chapter 9 test is below. The total is 10 points. Divide the student's score by the total of 10 to get a decimal number, and change that decimal to percent to get the student's percentage score.

Question	Max. points	Student score
1	4 points	
2	2 points	

Question	Max. points	Student score
3	4 points	
Total	10 points	

1. a. $1.20   b. $5.36   c. $5.71   d. $2.30

2. a. $0.12   b. $2.15

3. a. $1.56 + 1.56 + 0.78 = \$3.90$
   b. $2.55 + 2.55 = \$5.10$

Chapter 10 Test

Grading

My suggestion for grading the chapter 10 test is below. The total is 26 points. Divide the student's score by the total of 26 to get a decimal number, and change that decimal to percent to get the student's percentage score.

Question	Max. points	Student score
1	6 points	
2	2 points	
3	2 points	

Question	Max. points	Student score
4	4 points	
5	12 points	
Total	26 points	

1. a.
    $6 \times 1 = 6$

   b.
    $2 \times 7 = 14$

   c.
    $3 \times 3 = 9$

2. a. 4×6 b. 3×50

3. a. $8 + 8$ b. $3 + 3 + 3 + 3 + 3$

4. a.

   b.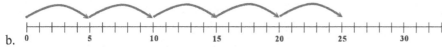

5. a. 12, 30   b. 0, 40   c. 6, 18   d. 12, 16   e. 24, 15   f. 40, 800

End-of-the-Year Test Grade 2 Answer Key

Instructions to the teacher:

My suggestion for grading is below. The total is 134 points. A score of 107 points is 80%.

Question	Max. points	Student score
Basic Addition and Subtraction Facts		
1	16 points	
2	16 points	
3	6 points	
	subtotal	/ 38
Mental Addition and Subtraction with Two-Digit Numbers and Word Problems		
4	1 point	
5	2 points	
6	3 points	
7	1 point	
8	3 points	
9	3 points	
10	6 points	
	subtotal	/ 19
Three-Digit Numbers		
11	2 points	
12	2 points	
13	2 points	
14	6 points	
15	4 points	
	subtotal	/ 16
Regrouping in Addition and Subtraction, including Word Problems		
16	3 points	
17	4 points	
18	2 points	
19	2 points	
20	2 points	
21	3 points	
	subtotal	/ 16

Question	Max. points	Student score
Clock		
22	6 points	
23	5 points	
	subtotal	/ 11
Money		
24	2 points	
25	2 points	
26	2 points	
	subtotal	/ 6
Geometry and Measuring		
27	2 points	
28	4 points	
29	1 point	
30	4 points	
	subtotal	/ 11
Fractions		
31	4 points	
32	6 points	
	subtotal	/ 10
Concept of Multiplication		
33	2 points	
34	2 points	
35	3 points	
	subtotal	/ 7
	TOTAL	/ 134

1. a. 13, 18, 11, 15   b. 11, 13, 12, 12   c. 16, 12, 11, 12   d. 14, 14, 14, 17

2. a. 9, 8, 7, 4   b. 8, 4, 8, 8   c. 9, 8, 9, 9   d. 9, 9, 6, 7

3.

a. $2 + 9 = 11$	b. $8 + 9 = 17$	c. $5 + 7 = 12$
$9 + 2 = 11$	$9 + 8 = 17$	$7 + 5 = 12$
$11 - 2 = 9$	$17 - 8 = 9$	$12 - 7 = 5$
$11 - 9 = 2$	$17 - 9 = 8$	$12 - 5 = 7$

4. $35 + 35 = 70$

5. $5 + 9 = 14$ They each got seven apples.

6. 10, 12, 14, 16, 18, 20

7. $90 - 75 = 15$

8. $32 - 16 - 10 = 6$ He needs six more dollars.

9. a. 8   b. 8   c. 7

10 a. 67, 70   b. 92, 54   c. 26, 31

11. a. 267   b. 908

12. 568, 578, 588, 598, 608, 618, 628

13. a. 417, 447, 714   b. 89, 809, 998

14. a. 600, 960   b. 500, 320   c. 332, 62

15. a. $92 = 92$ b. $248 < 824$ c. $170 > 125$ d. $400 < 404$

16. a. 83   b. 682   c. 748

17. a. 24, 61   b. 722, 970

18. The total cost was $304. $152 + 152 = 304$

19. $450 - 126 = 324$ DVDs

20. $218 + 218 = 436$

21. a.

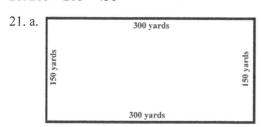

b. $300 + 300 + 150 + 150 = 900$ yards

22.

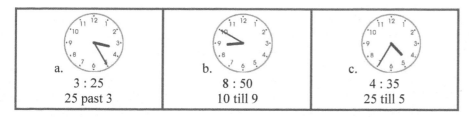

a. 3 : 25	b. 8 : 50	c. 4 : 35
25 past 3	10 till 9	25 till 5

23.

from	3:00	2:00	1 AM	11 AM	8 PM
to	3:05	2:30	8 AM	1 PM	midnight
amount of time	5 minutes	30 minutes	7 hours	2 hours	4 hours

24. a. $1.30   b. $5.41

25. The change is $0.65.

26. His change was $0.15.

27. Shape A: <u>a square</u>   Shape B: <u>a pentagon</u>

28. a.

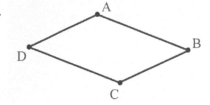

    b. a quadrilateral (also acceptable: a parallelogram)

    c. Side AB: about <u>about 2 inches</u>   Side BC: <u>about 1 1/2 inches</u>

      Side CD: about <u>about 2 inches</u>   Side DA: about <u>about 1 1/2 inches</u>

29. about 9 cm

30. It is enough for the student to mention one suitable measuring unit. All the units mentioned below are right answers.

Distance	Unit(s)
how long my pencil is	cm, in
the distance from London to New York	km, mi
the height of a wall	m, ft
the distance it is to the neighbor's house	m, ft

31.

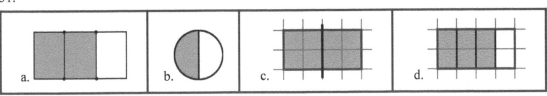

32.

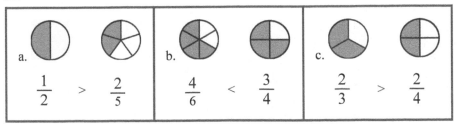

33. a. 2 × 3 = 6   b. 5 × 2 = 10

34. a. 3 × 5 = 15   b. 5 × 4 = 20

35. a. 10   b. 9   c. 30

Cumulative Reviews
Answer Keys

Cumulative Reviews Answer Key, Grade 2

Cumulative Review: Chapters 1 - 2

1. a. 8 hours   b. 24   c. $9.00 now   d. $40.00

2.

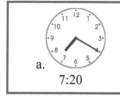 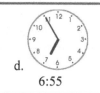

10 + 10 = 20	
15 + 15 = 30	
20 + 20 = 40	
25 + 25 = 50	
30 + 30 = 60	
35 + 35 = 70	
40 + 40 = 80	

3.

a.	b.	c.	d.
7:20	5:35	1:15	6:55

4. a. 3:40 and 4:15
   b. Jim walked his dog for one hour.
   c. 5:30 AM

5.

a. 3 + 6 = 9	b. 4 + 6 = 10	c. 3 + 5 = 8
6 + 3 = 9	6 + 4 = 10	5 + 3 = 8
9 − 3 = 6	10 − 4 = 6	8 − 5 = 3
9 − 6 = 3	10 − 6 = 4	8 − 3 = 5

6. A - HORSE

Cumulative Review: Chapters 1 - 3

1. a. 13, 17   b. 16, 13   c. 9, 8   d. 7, 7

2.

from	9 AM	6 AM	11 AM	12 AM	10 AM
to	1 PM	8 PM	4 PM	12 PM	2 PM
hours	4	14	5	12	4

3. a. five Tuesdays   b. Jan 20

4. a. She started at 12 noon.
   b. Grandma sleeps six hours.

5. a. 97, 34   b. 68, 23   c. 94, 38

6.

a. $18 + 4 = 22$	b. $75 + 5 = 80$	c. $56 + 3 = 59$
d. The difference of 8 and 12 is 4.	e. The difference of 43 and 49 is 6.	f. The difference of 21 and 30 is 9.

7. a. 5, 4   b. 5, 10   c. 3, 3

8.

a. $8 + 6 = 14$ $14 - 8 = 6$	b. $5 + 9 = 14$ $14 - 5 = 9$	c. $6 + 6 = 12$ $12 - 6 = 6$

9. a. 5, 8, 3   b. 6, 4, 6   c. 5, 5, 7   d. 8, 5, 6

10. $7 + 6 = 13$ $13 - 6 = 7$
    $6 + 7 = 13$ $13 - 7 = 6$

Cumulative Review: Chapters 1 - 4

1. a. There are fourteen boys in the class.
   b. Andy has $16. $8 + 8 = 16$
      Together they have $24. $16 + 8 = 24$

2. a. 32, 68   b. 48, 59   c. 53, 28

3. The camera costs $134 now. $67 + 67 = 134

4. a. 25 past 7   b. 10 past 5   c. 10 till 6   d. 20 till 1   e. half past 12   f. 11 o'clock

5. a. 81   b. 35   c. 81   d. 120

6. a. It took Chris 60 minutes to make the cards.
   b. He finished at 1 o'clock.

7. a. 32   b. 9   c. 8   d. 11   e. 27   f. 36

8. The total cost is $18 + $18 + $18 + $25 = $79.

1.

a. [T/T] and [TT/TT]	b. [T][T]/●●●● and ●●●/●●●/●●●
$20 + 40 = 60$	$24 + 7 = 31$
$40 + 20 = 60$	$7 + 24 = 31$
$60 - 40 = 20$	$31 - 7 = 24$
$60 - 20 = 40$	$31 - 24 = 7$

2. a. 119   b. 139   c. 107   d. 78

3. a. LLAMA   b. ORDINAL

4.

	a. 2:55	b. 7:10	c. 2:25	d. 11:00
5 min. later →	3:00	7:15	2:30	11:05

5. There are six blue flowers. $15 - 4 - 5 = 6$

6. a. $16 < 17$ b. $22 > 21$ c. $8 < 16$ d. $42 < 43$ e. $34 > 31$ f. $7 < 8$

7.

a. $36 + 22$	b. $72 + 18$
30 + 20 + 6 + 2	70 + 10 + 2 + 8
50 + 8 = 58	80 + 10 = 90
c. $54 + 37$	d. $24 + 55$
80 + 11 = 91	70 + 9 = 79

8. a.

464	474	484	494	504	514	524	534	544

b.

400	450	500	550	600	650	700	750	800

1. a. first b. fourth c. fourth d. second

2. He spends two hours practicing.

3. The total cost is $29. $12 + 12 + 5 = 29$

4. a. 98 b. 145 c. 143 d. 88 e. 82

5.

a.	b.	c.
$66 + 4 = 70$ $92 + 8 = 100$	$31 + 3 + 6 = 40$ $63 + 2 + 5 = 70$	$47 + 2 + 1 = 50$ $32 + 6 + 2 = 40$

6.

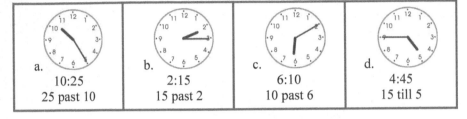

a.	b.	c.	d.
10:25 25 past 10	2:15 15 past 2	6:10 10 past 6	4:45 15 till 5

7. a hexagon:

8.

a. $1 = \dfrac{4}{4}$ b. $1 = \dfrac{6}{6}$ c. $1 = \dfrac{5}{5}$

9.

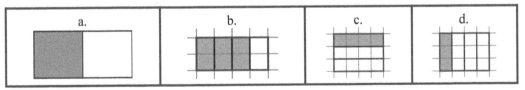

Cumulative Review: Chapters 1 - 7

1. a. She needs to buy six more apples. $8 + 8 - 10 = 6$
   b. Two go-karts would cost $36. $18 + 18 = 36$
   c. She has eight cats and there are four kittens. $4 + 4 = 8$

2.

    a. 1:25   b. 3:15   c. 15 till 3

3.

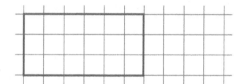

a. $577 - 10 = 567$	b. $926 - 0 = 926$
$577 - 20 = 557$	$926 - 100 = 826$
$577 - 30 = 547$	$926 - 200 = 726$
$577 - 40 = 537$	$926 - 300 = 626$
$577 - 50 = 527$	$926 - 400 = 526$
$577 - 60 = 517$	$926 - 500 = 426$

4. a. 17, 11, 14   b. 16, 12, 16   c. 13, 11, 14

5. a. a pentagon   b. a hexagon   c. a triangle

6. There are 18 squares in the grid.

7.

Ordinal Number	Name		Ordinal Number	Name
1st	first		8th	eighth
2nd	second		9th	ninth
3rd	third		10th	tenth
4th	fourth		11th	eleventh
5th	fifth		12th	twelfth
6th	sixth		13th	thirteenth
7th	seventh		14th	fourteenth

1. a. 11   b. 22   c. 37   d. 37

2. a. 9, 16   b. 12, 15   c. 14, 15   d. 18, 13

3.

a. $17 - 11 = 6$ Think: $11 + 6 = 17$	b. $43 - 37 = 6$ Think: $37 + 6 = 43$	c. $66 - 59 = 7$ Think: $59 + 7 = 66$
d. $35 - 28 = 7$	e. $80 - 77 = 3$	f. $100 - 94 = 6$

4.

$$- \boxed{4} \quad - \boxed{4} \quad - \boxed{3} \quad - \boxed{3} \quad - \boxed{6} \quad - \boxed{2} \quad - \boxed{5}$$

79   75   71   68   65   59   57   52

5.

a. 15 past 6 6:15	b. 20 till 3 2:40	c. 5 past 10 10:05	d. half past 3 3:30
e. 15 till 8 7:45	f. 20 till 12 11:40	g. 5 till 1 12:55	h. 25 past 1 1:25

6. a. Dan's sister weighs 118 pounds. $138 - 20 = 118$
   b. Together they weigh 256 pounds. $138 + 118 = 256$

7. a - b.

   c. triangles

8. a. $\frac{2}{6}$ $\frac{2}{3}$ $\frac{2}{4}$ $\frac{2}{5}$

   b. 2/3 has the most pie to eat.

9. a. <   b. >   c. $789 < 798$

1.

a. $\dfrac{1}{2}$ b. $\dfrac{1}{4}$ c. $\dfrac{3}{4}$ d. $\dfrac{2}{4}$ e. $\dfrac{4}{4}$ f. $\dfrac{2}{2}$

2. a. 5 hours   b. 7 hours   c. 5 hours   d. 4 hours   e. 4 hours   f. 4 hours

3. a. 15, 13, 6   b. 5, 6, 7   c. 7, 5, 7

4. a. $0.94   b. $3.14   c. $8.22

5. a. It has five corners.   b. a pentagon

6. He has 10 marbles. $15 - 5 = 10$
   Together they have 25 marbles. $15 + 10 = 25$

7. Eva ate six pieces of pie. $4 + 2 = 6$

8. a. 537, 807   b. 600, 240   c. 532, 352

9. a. 332, 670   b. 270, 541

10.

a. 79¢
Customer gives $1

Change: $0.21

b. $1.89
Customer gives $5

Change: $3.11

1.

1	2	3	4	5	6	7	8	9	10
11	12	13	14	15	16	17	18	19	20
21	22	23	24	25	26	27	28	29	30

2. a. I will return January 28.
   b. 2, 9, 16, 23, 30

3. She now has $20.

4. a. 729   b. 440   c. 191   d. 524

5. a. 14   b. 60   c. 30   d. 60   e. 140   f. 60

6. a. $0.71   b. $2.07   c. $9.11

7. Her change was $3.52 $5.00 - 1.48 = 3.52$

8. The total is 100 cents or one dollar.

9. Answers will vary.

10. a. No, she still needs four more dollars. $45 - 28 - 13 = \$4$
    b. The total cost of three rakes is $51. $17 + 17 + 17 = \$51$
    c. Four sacks of potatoes would weigh 88 kilograms. $22 + 22 + 22 + 22 = 88$